Renewable Electricity and the Grid

The Challenge of Variability

Edited by Godfrey Boyle

JUBILEE
CAMPUS
LRC

earthscan

publishing for a sustainable future

London • Sterling, VA

First published by Earthscan in the UK and USA in 2007
Reprinted 2009
Paperback edition first published in 2009

ISBN: 978-1-84407-418-1 hardback
ISBN: 978-1-84407-789-2 paperback

Typeset by FiSH Books, Enfield, Middx.
Printed and bound in the UK by TJ International
Cover design by Nick Shah

For a full list of publications please contact:

Earthscan
Dunstan House, 14a St Cross Street
London, EC1N 8XA, UK
Tel: +44 (0)20 7841 1930
Fax: +44 (0)20 7242 1474
Email: earthinfo@earthscan.co.uk
Web: **www.earthscan.co.uk**

22883 Quicksilver Drive, Sterling, VA 20166-2012, USA

Earthscan publishes in association with the International Institute for
Environment and Development

A catalogue record for this book is available from the British Library

Library of Congress Cataloging-in-Publication Data

Renewable electricity and the grid: the challenge of variability/Edited by
Godfrey Boyle.
    p. cm.
  Includes index.
  ISBN-13: 978-1-84407-418-1 (hardback)
  ISBN-10: 84407-418-8 (hardback)
  1. Renewable energy sources. 2. Electric power production. I. Boyle,
Godfrey.
  TJ808R417 2007
  333. 793'2–dc22   2007016031
  1006053351

FSC
Mixed Sources
Product group from well-managed
forests and other controlled sources

Cert no. SGS-COC-2482
www.fsc.org
© 1996 Forest Stewardship Council

Contents

List of Figures and Tables

FIGURES

TABLES

List of Contributors

Professor Dennis Anderson is an emeritus professor of energy and environmental studies at Imperial College London. He was formerly the Energy Adviser of the World Bank, Chief Economist of Shell and an engineer in the electricity generation industry.

David Andrews is a chartered engineer, and has been Energy Manager with Wessex Water for 10 years. He has worked in the energy business for most of his professional life, specializing in small-scale power generation using engines.

Dr Mark Barrett joined University College London in 2006. He specializes in the technical and economic modelling of energy and transport systems and atmospheric emissions. He has worked for universities, industry, government and environmental organizations. He has his own environmental consultancy, Sustainable Environment Consultants (SENCO).

Ümit Cali is an electrical/communication engineer with an MSc from the University of Kassel. He completed his BSc in electrical engineering at the Technical University of Yildiz-Istanbul in 2000. Currently, he is preparing his PhD on the wind power prediction systems. He worked in the field of telecommunication in IBM International & Sabanci Corp. in Istanbul. Since 2002, he has worked as a scientific staff member at ISET in the R&D division. His main research interests are wind power prediction systems, analysis of measured data and artificial intelligence techniques.

Bob Everett is lecturer in renewable energy at the Open University, Milton Keynes. He has an MA in mechanical and electrical engineering from Cambridge University and studied passive solar gains in houses for his PhD from the Open University. He has a background in electrical and computer modelling, and a research interest in combined heat and power generation of all sizes.

Dr Gregor Giebel has worked in wind energy since he started at Risø 11 years ago, and is now a senior scientist. His PhD, obtained from Oldenburg University, was about large-scale integration of wind energy and short-term prediction. Since then, he has worked mainly with short-term prediction of wind farm output and condition monitoring. He was the coordinator of the EU-funded CleverFarm project. In large-scale integration he assisted the WILMAR and TradeWind projects with knowledge on short-term forecasting. His main activity now is short-term forecasting, on a commercial basis and in several national and international research projects. Currently, he is the coordinator of the EU funded Coordination Action POW'WOW (Prediction Of Waves, Wakes and Offshore Wind).

Professor Tim Green gained BSc (Eng) (1986, Imperial College) and PhD degrees (1990, Heriot-Watt University) in electrical engineering. He is a professor of electrical power engineering and deputy head of the Control and Power Research Group and has been at Imperial College London since 1994. His research interests cover the assessment and integration of renewable energy sources into electrical networks, distributed generation, power quality and supply quality improvement with distributed generation, the control of flexible AC transmission systems and FACTS component development. He is a member of the IEE, senior member of the IEEE and a member of Cigre working group 14 on Voltage Sags and Power Quality.

Robert Gross MSc, MEI, is a lecturer at Centre for Environmental Policy, Imperial College London (ICEPT). He is head of Technology and Policy Assessment at the UK Energy Research Centre. He co-convenes Imperial's MSc in Environmental Technology, Energy Policy option. He has undertaken research and consultancy for diverse organizations and companies, from BP and several UK utilities to the United Nations Development Programme (UNDP), World Bank, the British Wind Energy Association (BWEA) and Greenpeace.

Philip Heptonstall gained his first degree in economics. After a 15-year career as a business analyst and project manager in the London wholesale insurance markets, he gained an MSc in environmental technology, specializing in energy policy, at Imperial College. He is currently a research associate with the UK Energy Research Centre and ICEPT, and is the lead researcher on the UKERC's Technology and Policy Assessment function.

Paul Hughes is now involved with mathematical modelling in the financial services sector. He previously worked with Airtricity for several years as a wind analyst. He originally graduated with a BSc in experimental physics and biology from the National University of Ireland, Maynooth. He followed that with an MSc in computational astrophysics from University College Dublin.

Brian Hurley has worked in wind technology since the early 1970s, while lecturing in engineering with the Dublin Institute of Technology. He is one of four founding members of Future Wind Partnership, which subsequently became Airtricity, a green electricity utility. He is currently Chief Scientist at Airtricity.

David Infield is professor of renewable energy systems with the Department of Electronic and Electrical Engineering at Loughborough University. He has worked in the renewables sector since completing his PhD in 1978. During this time his research interests have ranged over solar thermal system design, wind energy systems and photovoltaics. A common theme has been the integration of renewable energy sources and in particular electricity supply systems.

René Jursa is a physicist with an MSc from the University of Bremen and has been working at the R&D division 'Information and Energy Economy' at ISET since 2002. He is a scientific researcher in the area of wind power prediction. After his diploma he worked at the University of Bremen in the area of theoretical physics and following that in a company in the area of bioinformatics. His main scientific interests are neural networks, optimization algorithms and methods of signal processing.

Dr Bernhard Lange is head of Information and Prediction Systems in the R&D division, 'Information and Energy Economy' at Institut für Solare Energieversorgungstechnik (ISET). He is a physicist with an MSc from the University of Oldenburg. After graduating he worked in Denmark with Risø National Laboratory and Wind World A/S. From 1998 to 2002 he prepared his PhD on offshore wind power meteorology at Risø National Laboratory and the University of Oldenburg. For the last 10 years, his main research interests have been wind power meteorology, wind farm modelling and wind power forecasting.

Michael Laughton FREng, is Emeritus Professor of Electrical Engineering of the University of London with special interests in electrical power systems analysis. He is a co-editor of the *Electrical Engineer's Reference Book*, 14th to 16th editions (Newnes) and has acted as Specialist Adviser to UK Parliamentary Committees in both upper and lower Houses on alternative and renewable energy technologies and on energy efficiency. He is the UK representative on the Energy Committee of the European National Academies of Engineering, a member of energy and environment policy advisory groups of the Royal Academy of Engineering, the Royal Society and the Institution of Electrical Engineers as well as the Power Industry Division Board of the Institution of Mechanical Engineers.

Dr Matthew Leach is senior lecturer and deputy director of ICEPT. He leads research related to both decentralized energy systems and urban sustainability and directs an MSc stream in energy and environment. Originally an engineer by first degree, Matthew's research includes publicly and industrially funded research into energy and urban systems, combining mixtures of technical, economic, environmental science and policy analysis. As part of the 2003 Energy White Paper process, he worked with AEA Technologies for DTI on low carbon energy scenario modelling and led a follow-up DTI study reviewing international scenarios for low carbon technology costs. He led a review of low carbon technologies for the Prime Minister's Strategy Unit, and cogeneration work within the DTI Renewables Innovation Review. He is immediate past Chair of Council of the British Institute of Energy Economics and Vice-President of the Energy Institute.

David Milborrow is an independent consultant who has been involved in renewable energy for nearly 30 years. He originally worked for the Central Electricity Generating Board, where he managed one of the early integration studies and was also closely involved with moving forward some of the early windfarms in the UK. He has been a freelance consultant since 1992 and now specializes in economic and technical issues associated with the renewable energy sources – especially wind. He acts as technical adviser to a number of organizations, writes for the journal *Wind Power Monthly* and lectures at a number of universities on integration issues, offshore wind and wind energy economics.

Dr Kurt Rohrig is head of ISET's R&D division 'Information and Energy Economy'. Dr Rohrig has worked with ISET since 1991 and has been the scientist-in-charge for projects handling the online monitoring and prediction of wind power for large supply areas – carried out in cooperation with large power transmission utilities. The computer models and approaches developed in the frame of his work are in operation at all German transmission system operators with high wind power penetration. Furthermore, Dr Rohrig is head of the thematic network 'Energy and Communication', which consists of 12 partners from industry, universities and research institutes.

Florian Schlögl is a mechanical engineer with a diploma in software engineering from the University of Kassel. After two years in Sweden he started at ISET in 2002 in the R&D division 'Information and Energy Economy'. He deals mainly with the development of software tools for online monitoring and prediction of wind power to be used at transmission system operation centres.

Graham Sinden carries out research into the long-term characteristics of renewable electricity sources such as wind, wave, tidal power, the relationship between resource availability and electricity demand, and the integration of renewable electricity sources into larger electricity networks. He has previously worked in environmental policy and environmental science for the Environment Protection Authority (Australia). Graham is currently based at the Environmental Change Institute, Oxford University, and he holds a public position with the UK Department of Trade and Industry's Renewables Advisory Board.

Professor Jim Skea OBE, FRSA, is Research Director of the UK Energy Research Centre (UKERC). Before setting up UKERC in October 2004, he spent six years as Director of the Policy Studies Institute. He was instrumental in launching the Low Carbon Vehicle Partnership (LowCVP), an action and advisory group bringing together industry, academia, non-governmental organizations (NGOs) and government departments. He was previously Director of the Economic and Social Research Council's Global Environmental Change Programme and a Professorial Fellow at Science and Technology Policy Research Unit (SPRU), at the University of Sussex. His main research interests are: energy/environmental policies; sustainable development; climate change; environmental regulation and technical change; and general business and environment issues. He also chairs the Scottish Power Green Energy Trust.

Dr Fred Starr is presently working as a visiting scientist at the DG-JRC Institute for Energy at Petten in the Netherlands, which is one of the European Commission's research sites. He is currently advising the Institute on the design of plants to produce hydrogen and electricity from coal, and also on the EU Cogeneration Directive. He previously worked for 30 years with British Gas, where he initiated the Stirling Engine micro-CHP programme. More recently he has worked at ETD Ltd, a consultancy company specializing in high temperature materials, where he reviewed the problems of cycling of steam and CCGT plants.

Dr Simon Watson has a degree in physics from Imperial College, London and a PhD from Edinburgh University. His research career began at the Rutherford Appleton Laboratory involving wind resource assessment, wind speed forecasting and wind power integration. In 1999, Simon Watson moved to the green electricity supply company Good Energy. He joined the Centre for Renewable Energy Systems Technology (CREST) at Loughborough University in 2001 and is presently the Acting Director of CREST. Recent research interests include onshore, offshore and urban wind speed prediction, condition monitoring of wind turbines and the impact of climate change on the electricity supply industry.

Preface

The variability of power output exhibited by many renewable electricity sources represents something of a challenge to maintaining secure supplies in the integrated electricity systems of industrialized countries – especially if, as widely anticipated, the contribution of renewables to national grids rises to very substantial levels. But is this a major – or even an insuperable – challenge, or one that is readily amenable to solution? This is the key question this book attempts to address. It also raises a host of other important issues.

How do electricity systems *currently* cope with the hourly, daily and seasonal variability of demand, and with the sudden interruptions to supply that occasionally occur, for example due to the failure of a major conventional power plant? Are renewable sources 'intermittent', or is it more accurate to describe them as 'variable'?

How much 'dispatchable' generating capacity (fossil-fuelled or renewable) is required to provide supplementary 'backup' power for variable renewables? And how should 'backup' be defined? Is it principally for *power*, to provide short-term control of grid frequency; or is it principally for *energy*, to contribute to annual supply or long-term supply reserve requirements? To what extent can variable renewables such as wind contribute to 'firm power' and be accorded a 'capacity credit'; and how does such capacity credit vary with the proportion of renewables in the system?

How much backup capacity already exists on conventional electricity systems, and how much more will need to be added as the proportion of variable renewables increases? How much are these backup supplies likely to cost?

What *kinds* of backup supplies will be required in the renewables-intensive electricity systems of the future? To what extent can existing backup supplies be adapted to cope with a greater contribution from renewables? Are new, more flexible, forms of generation required? And what role might there be for various *storage* technologies, both existing and emerging?

What is the potential role of wind power *forecasting*, over the short and medium term, in enabling electricity system operators to adjust the output of other power sources to match the variability of wind? To what extent can the variability of wind be mitigated in future by contributions from other renewables, such as wave or tidal power? And to what extent could wide geographical dispersal, across the UK and over the rest of Europe, reduce the overall variability of renewable sources such as wind?

Should electricity demand management play a more important role in the 'informated' electricity grids of the future? And is it possible to envisage a future electricity system in which an extremely high proportion – perhaps 95 per cent – of electricity comes from renewables?

These questions, and many more, were originally addressed by the contributors to this book at a major conference held at the Open University in 2006. They have each expanded their conference presentations to produce the detailed and thoughtful analyses that follow. As might be expected, there are occasional differences of view, and of emphasis, between authors, and some areas of controversy remain to be resolved. But there appears to be a broad consensus that the variability of renewables is a not a substantial problem at present levels of grid penetration; and that, as the proportion of renewable generation rises in future, the problems that arise should be amenable to solution, as electricity grids, generating technologies and load management techniques evolve into more flexible, more sophisticated and more sustainable forms.

Godfrey Boyle
The Open University
July 2007

Acknowledgements

Thanks are due to my Open University colleagues Claire Emburey, Angie Swain, Toni Thomas, Bob Everett and David Elliott for helping to organize the conference that led to this book; and to Sally Boyle for redesigning some of the figures and for her help and support over many years.

List of Acronyms and Abbreviations

AC	alternating current
AEY	annual energy yield
AGR	Advanced Gas-cooled Reactor
AI	artificial intelligence
ANN	artificial neural network
BETTA	British Electricity Trading and Transmission Arrangements
°C	degrees Celsius
CCGT	combined-cycle gas turbine
CEGB	Central Electricity Generating Board (UK)
CF	capacity factor
CH_4	methane
CHP	combined heat and power
CO	carbon monoxide
CO_2	carbon dioxide
DC	direct current
DDC	Dynamic Demand control
DTI	UK Department of Trade and Industry
DWD	Deutscher Wetterdienst (German Weather Service)
EETS	Energy Efficiency Trading Scheme
EU	European Union
EU-15	15 European Union member states
EWEA	European Wind Energy Association
GBSO	Great Britain System Operator
GJ	gigajoule
GW	gigawatt
GWe	gigawatts of electricity
GWh	gigawatt hours
H_2S	hydrogen sulphide
HRSG	heat recovery steam generator
HVDC	high voltage direct current
Hz	hertz
IGCC	integrated gasification combined cycle
ISET	Institut für Solare Energieversorgungstechnik
km^2	square kilometres
$km\ h^{-1}$	kilometres per hour
kV	kilovolt
kW	kilowatt
kWh	kilowatt hour

kW m^{-1}	kilowatts per metre
LAM	local area model
LNG	liquid natural gas
LOLE	loss of load expectation
LOLP	loss of load probability
m s^{-1}	metres per second
MAE	mean average error
Mb	millibars
ME	mixture of experts
MOS	model output statistics
MSL	mean sea level
MSW	municipal solid waste
Mt	megatonnes
MW	megawatt
MWh	megawatt hours
MWkm	megawatt kilometre
National Grid	National Grid Electricity Transmission plc
NETA	New Electricity Trading Arrangements (UK)
NGC	National Grid Company
NGT	National Grid Transco (UK)
NNS	nearest-neighbour search
NWP	numerical weather prediction
OCGT	open-cycle gas turbine
OPEC	Organization of the Petroleum Exporting Countries
PIU	UK Cabinet Office's Performance and Innovation Unit
PJ	petajoule
PSO	particle swarm optimization
PV	photovoltaic(s)
PWR	Pressurized Water Reactor
r	correlation coefficient
REISI	Renewable Energy Information System on Internet
ROC	Renewable Obligation Certificate
RMSE	root mean square error
rpm	revolutions per minute
SCAR report	System Costs of Additional Renewables report
SNG	substitute natural gas
SPLD	system peak load demand
SVM	support-vector machine
TSO	Transmission System Operator
TWh	terawatt hours
TWh y^{-1}	terawatt hours per year
UKERC	UK Energy Research Centre
UTC	coordinated universal time
WPD	Western Power Distribution (UK)
WPMS	Wind Power Management System

Variable Renewables and the Grid: An Overview

Michael Laughton

INTRODUCTION

In all national electricity supply systems, the power demand varies over the course of a day; there is a rise and fall every 24 hours, with usually a night-time minimum and a daily maximum. In order to assess the contribution that renewable or other sources of energy can make to electricity supply, the distinction between energy and power has to be kept clearly in mind. Whereas the commercial operation of each generation plant is measured against total energy delivered, in the UK the central grid control operated by the National Grid Electricity Transmission plc (National Grid), acting in its role as Great Britain System Operator (GBSO), has to ensure that the power generated (the rate of delivery of energy) balances the power demand at all times, otherwise the system fails.

Ensuring power supply security requires a deeper understanding of grid-related issues than those related to energy supply availability. Naturally varying renewable energy sources certainly provide secure quantities of energy when considered over, say, a year, but of themselves do not necessarily guarantee the secure delivery of power as and when needed. The significance of the separation of requirements for energy delivery and power delivery (which seems to escape many commentators and advocates in the energy field) gives rise to separate power supply-related questions, such as those concerning plant capacity, generation load factors, system capacity planning margins, probabilistic measures of system power supply security, and backup plant requirements.

These questions will be considered further in this chapter from the view-point of guaranteeing grid security of power supply. Although difficulties and constraints are highlighted, it is taken for granted that renewable energy forms an important component in future energy supplies for the electricity supply industry, the more so in the UK with increasing dependence upon imported gas and the future retirement of coal and nuclear stations. Problems raised, there-fore, should be seen as problems to be solved – in some cases by more research, and in others by the development of technology.

RENEWABLE ENERGY SOURCE VARIABILITY

Renewable energy sources

Table 1.1 lists the main renewable energy sources used for electrical power generation, along with the distinctive types of development for energy conversion and extraction. Apart from the production of heat or chemicals, the generation of electrical energy is the main purpose.

Table 1.1 *Renewable energy sources for electrical power generation*

Renewable resource
Municipal solid waste (MSW)
Hydro: • large scale • small scale
Wind: • onshore • offshore
Biofuels: • energy crops • forestry wastes • agricultural wastes
Wave: • shoreline • near shore • offshore
Tidal: • stream • barrage
Solar: • photovoltaic

Source: Tyndall Centre (2003)

From the viewpoint of a power system operator, some of the difficulties associated with renewable source variability affecting the delivery of electrical power are as follows:

- uncertainties in predictions of power available at any given time, leading to scheduling difficulties, although obviously the degrees of uncertainty vary with the length of forecasting horizon;
- magnitude of fluctuations in power output, where small fluctuations can be accommodated easily, but larger fluctuations require special countermeasures;

- speed of fluctuations, where slow changes in resource availability and, hence, power output are usually predictable, but fast changes are less so.

In addition, there are generating plant performance abilities to be considered, such as power conversion limits where generating plant can operate efficiently only within certain limits of energy availability.

Variability of power generation

The characteristics of the varying electrical power outputs obtained from these respective resources vary considerably.

Tidal energy captured either from tidal streams of water or by storage and subsequent release in barrages is the most predictable of variable renewable energy forms. Tidal cycles lasting just less than 12.5 hours each day allow generation on either the ebb tide or on both the ebb and flood tides. Generation on the ebb tide with additional pumping at high tide is a further option, as shown in Figure 1.1. Studies of a potential barrage in the Severn Estuary show generation is possible for five to six hours during spring tides and for about three hours during neap tides. Thus, a tidal barrage produces two totally predictable but intermittent blocks of energy each day, the size and timing of which follow the lunar cycle.

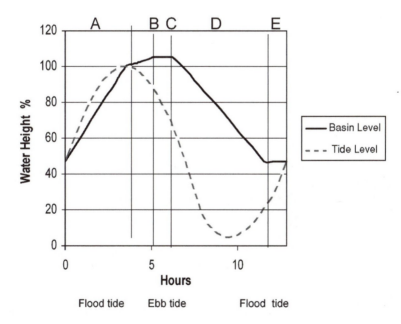

Note: A = filling; B = pumping; C = holding; D = generating; E = holding.
Source: Laughton (1990)

Figure 1.1 *Tidal cycle and electricity generation periods for a barrage with additional pumping at high tide*

Wave power is, at present, a relatively undeveloped and underemployed technology; therefore, without hard data the variability of power output from wave farms can only be surmised. It is known, however, that the variability depends upon both local and distant weather conditions. Wave power gives rise to further problems connected with plant limitations, apart from unpredictability. Figure 1.2 shows a typical probability relationship for wave power measured in kilowatts per metre (kW m⁻¹) of wave front for sea conditions in northern UK waters. Note the logarithmic scale. To design power take-off devices to capture the power in low-probability high-power waves would be too expensive; therefore, such devices are sized to cope with only a limited range of wave power levels that have a higher probability of occurrence. However, to withstand extreme conditions without being destroyed, the structure has to be designed to withstand such extreme events, regardless of their low probability.

Photovoltaic (PV): the power and energy output of any PV array depends upon the irradiance, which, in turn, depends upon the time of day and the time of year, the maximum power generated, and the length of operation achieved in summer. Local weather conditions result in individual array power outputs with many spikes and troughs, although the overall daily power output from several arrays spaced across the country should follow approximately a bell-shaped curve centred on midday, with a spread depending upon the length of daylight. In the UK, midsummer irradiance could last, for example, from 05.00hrs to 21.00hrs, but with levels falling from 100 per cent at 12.00hrs to 70 per cent at 16.30hrs, and then steeply to less than 20 per cent by 18.00hrs.

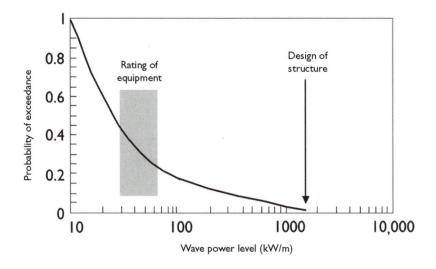

Source: Select Committee on the European Communities (1988)

Figure 1.2 *Typical annual variation in wave power levels*

PV energy generation, therefore, is in the form of one block of energy each day during daylight hours with power levels achieved being both seasonally and weather dependent.

Wind power, like wave power, requires the generating plant to withstand extreme conditions without being destroyed. Wind turbines are currently designed to withstand maximum wind speeds of usually around 25 metres per second (m s^{-1}), at which level the turbines are switched off for protection. Figure 1.3 illustrates a typical power output characteristic for a wind generator showing output rising from a cut-in wind speed of about 4m s^{-1} to 5m s^{-1}, to a maximum output at about 13m s^{-1} to 14m s^{-1} and a shut down speed at 25m s^{-1}.

Three sources of variability are apparent. First, there is zero output below cut-in wind speeds; second, between cut-in and maximum output, varying wind speeds can cause large changes in output, although these would tend to be smoothed out with many turbines covering a wide area; and, third, the turbine is switched off in storm conditions.

This last circumstance is illustrated in Figure 1.4, showing spot prices in the Danish electricity market during the first week of January 2005. Particularly strong winds during this time first of all produced ample supplies of wind power that sent the spot (marginal) prices to zero, followed by rapid rises in price as wind conditions strengthened beyond 25m s^{-1} wind speed and many wind turbines were shut down. Such fast fluctuations in output may be anticipated but are difficult to predict accurately, both in degree and in time, without knowledge of the extent and progress of the particular storm circumstances.

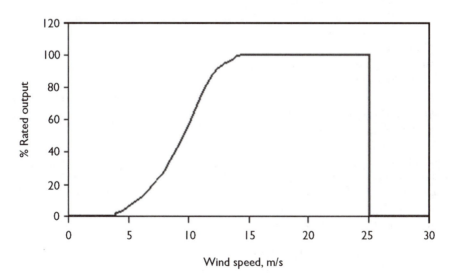

Source: Boyle (2004)

Figure 1.3 *Wind turbine output characteristics*

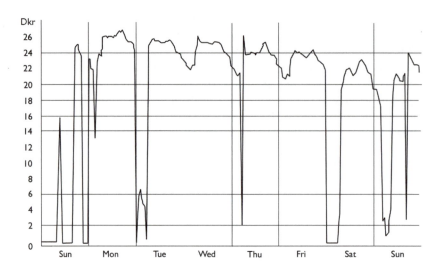

Note: DKr = Danish Kroner.
Source: www.nordpool.com/nordpool/spot/index.html

Figure 1.4 *Influence of storm conditions on spot electricity prices in Danish Kroners per megawatt hour in West Denmark during the first week of January 2005*

More usually, the total wind power output from a number of wind farms across a region is subject to slowly occurring large fluctuations caused by the changing regional weather patterns. Figure 1.5 shows such variability at the end of April and beginning of May 2004 in the E.ON Netz system in North Germany (E.ON Netz, 2004). Although the changes in wind power output represent some 80 per cent of installed capacity, such variations are more easily predicted than the sudden storm disconnections. The problem faced by E.ON Netz is more to do with the measures that need to be taken to ensure that system frequency is controlled and power flows in a coordinated manner across the transmission network.

Care should be taken in drawing parallels, however, between experiences in Germany and Denmark and the situation elsewhere, such as in the UK. Wind conditions over the whole British electricity supply system should be assumed to be different unless proved otherwise. Differences in latitude and longitude, the presence of oceans, as well as the area covered by the wind power generation industry make comparisons difficult. The British wind industry, for example, has a longer north–south footprint than in Denmark, while in Germany the wind farms have a strong east–west configuration. In addition, both Denmark and Germany operate a feed-in tariff system of support that allows wind generation to be determined entirely by the wind conditions and tariff levels – hence the uncontrolled Danish spot prices. Such support is common within the European Union (EU), but is not in the UK or

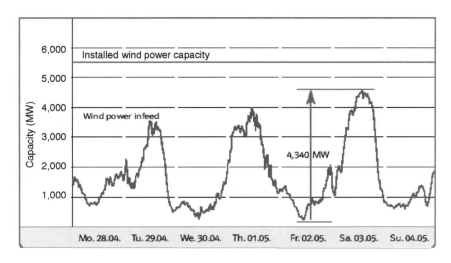

Source: E.ON Netz (2004)

Figure 1.5 *Large fluctuations in wind output in the E.ON Netz network in Germany*

in the Republic of Ireland, Belgium, Italy and Sweden, where renewable energy is supported through the use of regulated volumes.

A different and rather curious long-term variation in wind power, and one needing, perhaps, greater study, was illustrated at a Colloquium held in 1987 at the Electrical Engineering Department, Imperial College, London, on Economic and Operational Assessment of Intermittent Generation Sources on Power Systems. This variability relates not to days, weeks or seasons, but to observed changes in wind strengths over several decades. Figure 1.6 shows the cubed values of annual mean wind speeds at a wind recording station at Southport Marshside, which are proportional to wind turbine power output. The smoothed curve demonstrates the effect of applying a 15-term Gaussian filter to suppress the short-term fluctuations. Evidently, there are longer-term changing atmospheric circulation patterns present that modulate annual wind speeds (Palutikof and Watkins, 1987).

In this example, it is interesting to note that the energy output in a poor year would be less than half that in a high wind speed year; that high and low wind speed years also occur in spells; and that substantial variations about the average annual wind speeds can occur at all times. Perhaps of more concern is that such long-term mean annual wind speed variations may cause average power output – and, hence, average annual energy contributions – to differ substantially from predicted levels. This would be a matter of some consequence for both wind farm owners and system planners. More work needs to be done in defining these long-term cycles so that average annual wind conditions converted to hypothetical wind power over the last 50 years can be determined and anticipated in the future for, say, 2010, 2015, 2020 and 2025 – that is, during the lifetimes of present wind turbines.

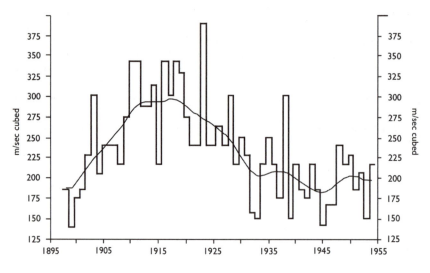

Source: Palutikof and Watkins (1987)

Figure 1.6 *Example of longer-term changes in average annual wind speeds*

Notwithstanding these long-term considerations, the effects of wind variability on the British grid have been studied in some detail over the past 30 years, and some of the conclusions are presented in the following sections.

Anticyclones and large-scale variability

The major questions concerning the relationships between variability of energy sources and grid power supplies occur when small-scale random local variability gives way over longer time periods to large-scale changes affecting much of the grid system supply from wind. Such large-scale variability would occur for wind and wave power when a substantial high-pressure weather system (anticyclone) moves in over the whole country or a large part of it and – with little or no wind – wind power output (and, potentially also, wave power generation) drops to near zero. The same large-scale intermittency would occur for PV generation at night or for the output of the Severn Barrage twice per day. Borrowing a descriptive term used in the analysis of electronic systems where a single reason is the cause of multiple circuit or equipment failures, this single phenomenon causing a general reduction of power generated from many geographically distributed sources may be termed *common mode failure* (Laughton, 2002).

Large anticyclones with little wind pass over the country throughout the year (see Wetterzentrale, undated) Those occurring in the winter are invariably accompanied by low temperatures, frost and, maybe, fog – the occasions when heating and lighting loads can also be at maximum (i.e. at winter peak load times). In the summer, similar conditions are invariably associated with clear skies and high temperatures when increased demand for cooling and air conditioning result.

An example of such winter conditions is shown in the Table 1.2 (Meteorological Office, 2006). The data summarizes the wind speeds, temperature conditions and atmospheric readings at some 66 Meteorological Office recording stations covering Scotland, north-west England, north-east England, Wales, the Midlands, East Anglia, south-east England and south-west England.

Although the north-west of Scotland is generally the windiest area of Britain, in this example the stronger winds, on average, were in the south of the country.

Table 1.2 *Winter anticyclone conditions on Wednesday, 28 December 2005, at 18:00 GMT*

Region and number of Meteorological Office stations	Wind speed range (knots)	Average wind speed (knots)	Temperature range (°C)	Average temperature (°C)	Pressure range at mean seal level (MSL) (mb)	Average pressure at MSL (mb)
Scotland (mainland) 13 stations	0 to 9	3	−7 to 3	−2	1020 to 1023	1022
North-west England 5 stations	0 to 9	3	−4 to −2	−3	1020 to 1021	1021
North-east England 6 stations	0 to 6	4	−3 to 0	−2	1020 to 1022	1021
Wales 6 stations	0 to 3	2	−5 to −1	−3	1020 to 1021	1021
Midlands 10 stations	0 to 9	6	−3 to 0	−1	1018 to 1021	1020
East Anglia 4 stations	7 to 11	9	−2 to 0	−1	1017 to 1018	1018
South-east England 13 stations	2 to 9	7	−3 to 1	−1	1017 to 1020	1019
South-west England 11 stations	0 to 9	5	−3 to 0	−2	1020 to 1021	1020

Note: Wind speeds are measured in knots. 1 knot = 1.15 miles per hour (mph) = 0.514m s^{-1}.
mb = millibars.
Source: www.metoffice.com/education/archive/uk

The average wind speed measured at all stations across the country is about 5 knots. Nevertheless, this is a somewhat meaningless measure: because of the non-linearity of wind turbine power output as shown in Figure 1.3, the total power generated depends mainly upon the number of sites with higher wind speeds. Here the relationship between the number of sites and wind speeds is shown in Figure 1.7.

It is important to note that although there are some stations recording zero wind speeds, overall there are winds across other parts of the country, albeit light winds. The zero wind condition affecting the whole country is a hypothetical possibility; but studies show that this is a very rare occurrence and may be ignored. For purposes of relating wind conditions to power generation, the light wind conditions provide the determining circumstances that render any consideration of even lower wind speeds irrelevant.

With regard to the typical power output/wind speed characteristic shown in Figure 1.3, for the purposes of the analysis here, wind turbines may be considered to start up at approximately 9 knots (4.63m s^{-1}), reaching full output somewhere above 22 knots and shutting down at approximately 50 knots.

According to these criteria, it would appear in the above example that only one station would be recording sufficient wind speed where power output could be obtained. It should be noted, however, that:

• Many Meteorological Office stations are geographically irrelevant for judging wind power potential in Great Britain (i.e. at airports instead of on hilltops).

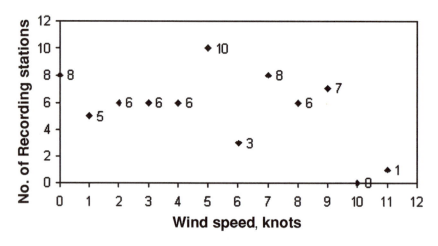

Source: www.metoffice.com/education/archive/uk

Figure 1.7 *Wind data for mainland Britain on Wednesday, 28 December 2005, at 18:00 GMT: Light winds with cold anticyclone weather*

- Meteorological Office data is collected at lower heights than hub heights; thus, wind speeds at Met Office station monitoring heights need to be increased to account for variation of speed with height.
- More complex rules have been developed; but a simple rule is:

$$V_z = V_h \, (z/h)^a \qquad\qquad\qquad [1]$$

where V_z and V_h are wind speeds at heights z and h, and where $h > z$, with a = 0.16 (Palutikof and Watkins, 1987).

Applying this height correction factor between a recording height of 30m and a hub height of 80m, recorded wind speeds of 9 knots could be about 11 knots at hub height, sufficient for a small amount of wind generation, but well below the maximum outputs achieved at over 20 knots.

GRID OPERATIONAL REQUIREMENTS

Power demand and supply: Daily load curves

In the British system covering England, Wales and Scotland, the total daily power demand varies between a minimum summer load of about 22.4 giga-watts (GW) and a winter peak above 59.4GW (National Grid, 2006). Figure 1.8 illustrates the demands during the two days, with minimum and maximum demands, respectively, over the 12-month period of 1 July 2005 to 30 June 2006, the areas under the curves representing daily energy demands/supplies and the heights of the curves representing the average half-hourly power demands. Here, the annual base load can be seen to be of the order of 22GW; but, with peak load occurring during the winter, the daily base loads would have higher values over the periods during the winter months.

To meet the continuously changing power demands, a mixture of different types of generation plant with varying degrees of responsiveness is needed to meet the base load, mid range and peak loads. Large predictable daily changes in demand or variable output from renewable generation plant are met by scheduling and contracting the conventional generation as appropriate. However, the load also fluctuates continuously in a random manner on a much smaller scale within a few percentage of the expected value. Generation of power, including all the variability of the power output of renewable sources, has to equal load plus losses at all times. As a result, balancing generation from spare plant has to be brought in and out of the system by the National Grid as required. Some of this spare capacity would be on 'hot standby' (i.e. connected to the network and operating at part load to ensure a stability of connection, as in the case of steam plant) or available for instant start-up and connection (as is the case for hydro, gas turbine and standby diesel plant).

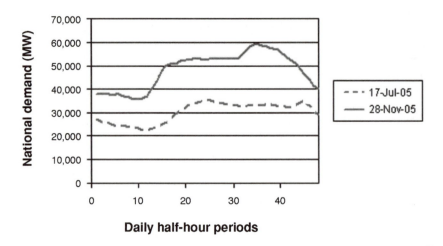

Source: www.nationalgrid.com/uk/Electricity/Data

Figure 1.8 *Daily load variation on the UK National Grid system showing maximum and minimum demand days from 1 July 2005 to 30 June 2006*

Power quality

The system frequency at 50 hertz (Hz) has to be controlled within the specified limits of ±0.5Hz by balancing power supply and demand. Exact balancing is not possible at all times, leading to changes in frequency as variations in the total energy flow are absorbed and returned by the large reservoir of energy afforded by the combined rotating masses of the large turbine-generating plant in the conventional power stations. This plant, locked together in synchronism, rotates at 3000 revolutions per minute (rpm) ±30rpm and provides the flexibility necessary for the instantaneous balancing of supply and demand. Unlike other interconnected continental power systems, the British system is an island system connected only by high voltage direct current (HVDC) links to neighbouring systems. HVDC links do not provide frequency control; thus, frequency control has to be exercised entirely within the national system.

Frequency control is but one aspect of guaranteeing power supply quality and is an important but seldom debated aspect of the power supply security question. Increasing levels of variable renewable generation capacity will bring both advantages and disadvantages for the maintenance of good power quality, although much remains to be learned in this respect. The importance of maintaining good power supply quality without voltage dips, surges, harmonics, frequency variations or interruptions in supply, even for milliseconds, does not feature in debates on the future of the industry; yet, without high-quality electrical power supplies, the operation of a modern industrialized society is not possible (Stones, 2003). Poor power quality can have large detrimental effects on industrial processes and in the commercial sector, with substantial costs associated with machine downtime, clean-up costs, product quality and equipment failure. Since this is a subject of specialist interest, however, it is not included for further discussion here.

Security of power supply and planning margins

Historically, in the development of national electricity supply industries, the planning process has sought to ensure that sufficient spare generation capacity is available, over and above that needed to meet the maximum peak load demand, in order to account for contingencies. To do so requires having sufficient spare capacity in order to meet not only expected generating plant outages for repair and maintenance, but also unexpected events causing breakdown of plant and, thus, non-availability of generating capacity.

This practice ensures that security of power supplies, as measured by loss-of-load probabilities, stays within the historic norms. It is also central to the determination of conventional generating plant capacity requirements where substantial variable renewable generation capacity is added to a system.

With variable renewable power supplies added to the system, all analyses show that wind generation plant, for example, can contribute to the security of supply to a certain extent, despite the arbitrary nature of the prevailing wind speeds. There is a suggested reliable capacity credit factor, determined by using known existing conventional plant reliability statistics combined with simulated wind turbine output based on measured meteorological office wind speed data. This capacity credit factor is simply an indication of the amount of existing conventional base-load capacity that could be displaced for various levels of wind penetration without degrading the overall system reliability standards. It is, therefore, also an important indication of the acceptable decrease in conventional capacity in the planning margin.

Before privatization of the UK electricity supply industry in 1989, the Central Electricity Generating Board (CEGB) used a planning margin of 24 per cent to provide generation security when planning the need for future generation installed capacity. After privatization, under the initial electricity 'pool' trading arrangements, which in the UK preceded the New Electricity Trading Arrangements (NETA), capacity payments were paid with regard to available generation capacity. These capacity payments, which were a function of loss of load probability (LOLP), were intended to provide a signal of capacity requirements. Under NETA – and now its successor, the British Electricity Trading and Transmission Arrangements (BETTA) – the plant capacity margins are currently determined solely by market forces.

The present planning margin in the UK advocated by the National Grid is now around 19 to 20 per cent at times of peak load, based on known conventional plant outage rates and the short construction times required by combined-cycle gas turbine (CCGT) stations. This margin equates to approximately 11.5GW, or a generator availability of about 71GW, at the time of peak load during the period of 1 July 2005 to 30 June 2006.

This spare capacity is, of course, a planning figure relating to a target for investment in capacity. The actual generation available and the margin over and above demand during the year, accounting for outages, vary considerably, as shown in Figure 1.9 (National Grid, 2006) and, in terms of the percentage margin relative to demand, in Figure 1.10. The difference between generation availability and national surplus accommodates other needs for capacity that might be called on, such as for control or contingency requirements.

The graphs show that although the peak demand of 59,632MW occurred in week 22 (21–27 November) when the generation available margin was 21 per cent, the minimum generation available margin of 5 per cent occurred in week 34, when the peak demand was 57,123MW. The consistency of the peak demand levels is also shown: the weekly peak demand did not fall below 57,000MW from week 18 to week 34.

BASE-LOAD CAPACITY DISPLACEMENT WITH INCREASING WIND PENETRATION

National wind characteristics and power generation

Studies of wind speeds in Great Britain show that there are significant periods in an average year when demand is high and wind output over the whole country is low. In particular, a typical year would have over 1600 hours when wind-generated output would be less than 10 per cent of maximum rated installed wind-generation capacity, including 450 hours when demand is between 70 to 100 per cent of peak demand (OXERA, 2003). Although the risk of system failure is greatest when demand is at its absolute peak, the risk is still significant for demands within a few percentage of the peak, say within 2GW to 4GW of peak in the system of the National Grid. Previous studies noted that thermal plant output may have a standard deviation of between 1GW and 2GW around the peak availability (Grubb, 1988).

Generally, in Great Britain stronger winds occur in winter; thus, during the winter season the average winter wind power available exceeds the mean annual level. Previous studies (Grubb, 1988) have shown, however, that the correlation with the peak 1 per cent of demand is negative, suggesting a tendency for the very highest levels of demand to be associated with less wind energy than might be expected. This lends support to the belief that the highest demands can (but not necessarily) occur on cold calm days (Grubb, 1988). The same conclusion can be drawn from Figure 1.11 from the data of hourly load factor and peak demand, derived from ten years of Great Britain electricity demand data and ten years of simulated wind-generation data, with each actual hour of wind speed matched against each actual hour of demand (OXERA, 2003). The graph shown in Figure 1.11 confirms that higher load factors (i.e. higher wind speeds) occur with higher percentage peak demands (i.e. winter demands). The significant section of the characteristic is in the droop shown in the load factors as peak demands approach the highest values. While not confirming that peak demands are always associated with the calm

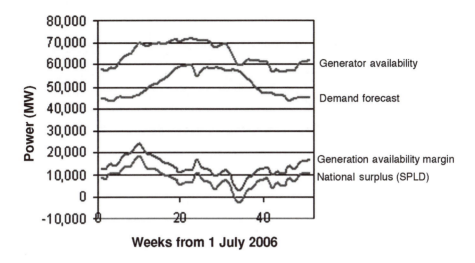

Figure 1.9 *National Grid demand and generation capacity available for the 12 months from 1 July 2005, with reference to the weekly system peak load demands (SPLDs)*

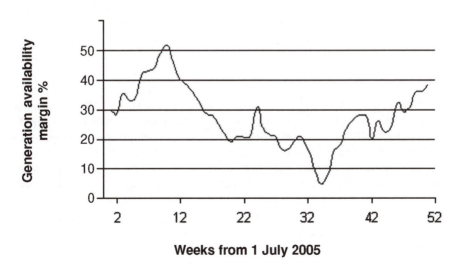

Figure 1.10 *Percentage generation availability margin relative to demand from 1 July 2005*

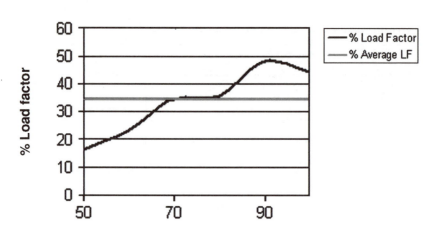

Source: OXERA (2003)

Figure 1.11 *Relationship between percentage of Great Britain peak demand and overall percentage hourly wind plant load factor*

cold weather accompanying winter anticyclones as illustrated in Table 1.2, the graph nevertheless indicates their increasing presence at such times.

Although attention is focused on variations in the overall national supply of wind power, it must not be forgotten that significant variations in supply could also come from more limited geographic areas if a large number of wind farms were contained therein. Significantly large concentrations of wind power generation capacity are being developed in Great Britain, both onshore and offshore; therefore, a two-year history of the output of a group of wind generators having rated capacities of 99MW in the Scottish Power Transmission area (southern Scotland) serves as a guide. These wind generators are geographically well distributed across the region; yet, between April 2003 and March 2005, for a total of 20.3 per cent of the half-hour periods, the aggregate output of all the wind farms was less than 5 per cent of the total capacity. Furthermore, for 12.6 per cent of the half hours, the output was less than 2 per cent of capacity, and in 2.2 per cent of the half hours there was no output at all (Bell et al, 2006). There is insufficient capacity in this example to have any significant impact on grid operations; nevertheless, the principle is clear that a region of the country with potential for considerable wind farm development can also experience large decreases in wind power output that should be evaluated in relation to the grid capacity requirements.

The relationship between hypothetical wind capacity and energy generated per annum for Great Britain can be seen in Figure 1.12. Again, a study of several years of hourly wind data gathered from Meteorological Office sites around the country, when processed through typical wind turbine power output versus wind speed characteristics, produced an annual probability

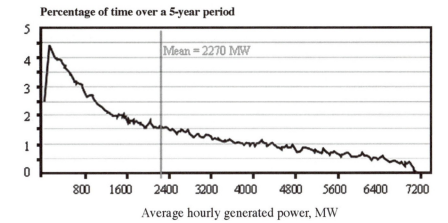

Source: National Grid (2002)

Figure 1.12 *Probability distribution of total Great Britain wind power generation from 7600MW of dispersed wind turbines*

distribution as shown (National Grid, 2002). Similar data distributions have been found for other countries.

Figure 1.12 shows the probability of achieving various power output levels from wind turbines over the whole country, given a theoretical total wind turbine-installed nameplate capacity of 7600MW. It is seen that the total average hourly power output calculated from Met Office wind data covering the country for every hour over the last five years can vary from 7300MW to practically zero. The mean output is 2270MW, which over a year would provide approximately 20 terawatt hours (TWh) of energy, or about 5 per cent of forecast national electrical energy demand in 2010, and would meet half of the government target for electrical energy generated from renewables.

Figure 1.12 is of fundamental importance in understanding the reasoning behind capacity credit estimates for various levels of wind penetration without resorting to mathematical explanations.

Wind capacity credit

In various studies (Grubb, 1986; Grubb, 1987; ILEX Energy Consulting, 2002) an historic security of supply standard of 9 per cent is commonly applied to the statistical probability of peak winter demand exceeding available supplies. Typically, these simulations of generation reliability use generic 500MW generating units with a probability of 85 per cent availability and wind power data obtained either from recorded wind speeds translated through manufacturers' power curves or, more accurately, from half-hourly metered generation from UK wind farms. Assuming no correlation between failures of conventional generating units, the behaviour of conventional plant and wind-generating plant is statistically combined, enabling the risk of peak demand exceeding available generation to be found. The minimum number of conventional generating units necessary to ensure that the loss of load probability, or risk of loss of supply, is less than 9 per cent is then determined; as a result, the capacity credit of wind is found (i.e. the ability of wind to displace conventional capacity for various levels of wind penetration).

Small capacity shortages have a much higher probability of occurring than large shortages, but have little effect on security of supply. The total wind power available in Great Britain over a short period of time (e.g. one to three hours) will vary randomly; but these variations are small, of the order of a few hundred megawatts, and are capable of being balanced out by the National Grid Company (NGC) using its existing controls and available plant. As the capacity of wind in the system increases, however, the consequences of occasional large decreases in wind output are of increasing concern.

Implications for conventional plant capacity needed

All study results indicate that for low levels of penetration, the firm power capacity displaced equals the mean power delivered by wind generation (i.e. measured by the total wind generation average load factor), but decreases with increasing penetration of wind (Rockingham, 1980; Halliday et al, 1983; Grubb, 1986, 1988; Swift-Hook, 1987). This diminishing return in the value of wind capacity reflects the increasing importance of the possibility of little output from all sources of wind generation.

This effect of a declining influence of wind on power system security of supply with increasing levels of wind-generation capacity installed is shown in Figure 1.13. These charts show the results of a further study by the National Grid combining the outputs illustrated in Figure 1.12 with National Grid operational models of conventional plant availability for various levels of wind capacity on the system. The peak demand shown here is 50,000MW; when repeated for 70,000MW, the study gave the same results.

The question posed in the study is: given a probability distribution of wind power generated nationally, how much conventional base-load plant capacity can be removed from the system without compromising system security, as measured by the LOLP criteria of not being able to meet demand more than ten years in a century?

Figure 1.13 shows power output probability distributions for 500MW, 7500MW and 25,000MW of hypothetical wind-generation capacity, respectively, spread around the system (National Grid, 2003). In these charts, zero represents generation balancing load. The areas under the curves to the left of the zero represent the probability of loss of supply where load demand exceeds generation capacity; the area to the right represents the probability representing generation capacity exceeding load demand. With the LOLP value chosen, the area to the left of zero is 9 per cent of the total area under the curve for any wind capacity chosen. In practice, the conventional plant capacity is adjusted so that the probability that demand exceeds available supply is 9 per cent (i.e. nine winters per century, the CEGB Generation Security Standard).

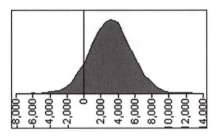

500 MW wind
59,000 MW conventional
Spare capacity=9.5GW

7500 MW wind
57,000 MW conventional
Spare capacity=14.5GW

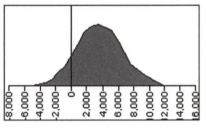

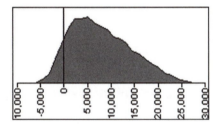

25,000 MW wind
55,000 MW conventional
Spare capacity=30GW

Note: Zero indicates generation balances load. The area to the left of zero is the probability of not meeting 50,000MW peak demand nine winters per century.
Source: National Grid (2003)

Figure 1.13 *Probability distributions of total generation capacity for secure supply*

The importance of the shape of the curve in Figure 1.12 is now apparent, with the left skewing towards lower values of power output prohibiting the replacement of a larger amount of conventional generation. Starting with a conventional capacity of 59,500MW, the conventional capacity displaced by wind generation is seen successively as 500MW, 2500MW and 4500MW: these are the capacity credit values calculated here for the penetration levels of wind of 500MW, 7500MW and 25,000MW, respectively, in the British National Grid system.

One important factor to note is the increase in the system's total capacity over and above that needed to meet the load. With only 500MW of wind installed, the excess capacity is 9.5GW, or 19 per cent of peak load, which is in line with the National Grid planning margin. With 25,000MW of wind capacity installed, however, which would provide some 16 per cent of the national demand for electrical energy, the excess capacity rises to 30GW, or 60 per cent of the peak load. How a liberalized market would accommodate the needs of financial return with the need for security of supply in such circumstances remains to be seen.

Other studies have produced the same results for the system in England and Wales (Grubb, 1987) and for Great Britain (ILEX, 2002). Collectively, the estimates of the increases of wind capacity credit with increasing wind capacity installed are shown in Figure 1.14. The curve showing Grubb's (1987) results represents a central value: his estimates deviate both above and below the solid curve depending upon different regional concentrations of wind farms.

To a first approximation for the British system:

$$\text{Wind capacity credit} = (\text{GW of wind capacity installed})^N \qquad [2]$$

where, for a central value, $N = 0.5$.
    Or:

$$\text{GW of capacity credit} = \sqrt{(\text{GW of wind installed})} \qquad [3]$$

or, with regional variations, $0.43 < N < 0.6$.

Demand growth scenarios with various penetration levels of wind energy by 2020

As already referred to, the problem to be faced in the future is how to accommodate high levels of variable wind capacity in a power supply system if security of supply considerations (i.e. capacity credit limitations) do not allow the release of alternative conventional generation capacity. This situation is illustrated in the results of a study for the UK Department of Trade and Industry (DTI) (ILEX Energy Consulting, 2002), in which future demands were postulated along with a high degree of penetration of wind power capacity. The results are shown in Table 1.3.

Table 1.3 *High electricity demand growth scenarios considered for Great Britain with various penetration levels of wind energy by 2020*

Peak demand (MW)	Energy from wind	Installed wind capacity (MW)	Conventional capacity* required (MW; margin)	Other renewable capacity (MW)	Excess capacity (MW)	Excess capacity margin
75,700	0%	0	90,083 = 19%	1600	15,983	21%
75,700	10%	9900	86,800 = 15%	1600	22,600	30%
75,700	20%	24,000	84,000 = 11%	1600	33,900	45%
75,700	30%	38,000	82,500 = 9%	1600	46,400	61%

Note: * Includes combined heat and power (CHP).

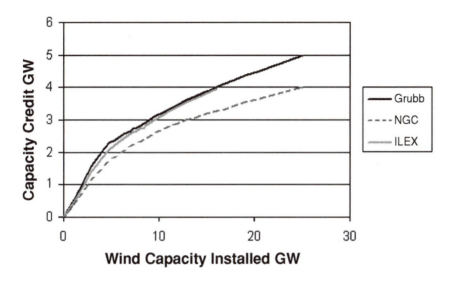

Figure 1.14 *Wind capacity credit in Great Britain relative to the National Grid security of supply standards*

The study assumed a fixed peak demand of 75.7GW on the Great Britain system by 2020 and a fixed non-wind renewable capacity of 1.6GW. It then examined the consequences, while maintaining reliability of supply, of adding to the system increasing amounts of wind-generation megawatt capacity. In this study, capacity remix was considered by adding support plant, such as open-cycle gas turbines (OCGTs), for purposes of maintaining adequate response and reserve requirements.

Here again the observation emerges from these results that irrespective of the accuracy of the assumptions concerning peak demand or energy required, nominal capacity margins increase dramatically (i.e. wind-generated electrical *energy* replaces other energy from other generators, but does not fully obviate the need for other generating *capacity*). Furthermore, the original conventional plant capacity planning margin of 19 per cent is never reduced to less than 9 per cent (i.e. this percentage of conventional plant capacity has to be maintained to exceed peak demand).

Backup capacity and security of supply

Unfortunately, the term 'backup capacity' has many meanings, and this has led to a great deal of misunderstanding (see Chapters 3 and 4 in this volume). It is applied, respectively, to both the need for capacities to support power requirements and the need for capacities to support energy requirements. Unfortunately, these two capacity needs are different, hence the muddle, especially since one is large and the other, if it exists at all, is small.

One major cause of some confusion is what to call the extra capacity in the system that has not been replaced by the added wind capacity, and which would run in parallel with wind generation. Various terms have been used and are still being applied to this retained capacity, such as 'backup capacity', which has caused perplexity (Laughton, 2002; Royal Academy of Engineering, 2002), or 'shadow capacity' (E.ON Netz, 2004). Other terms include 'compensating capacity' and 'balancing capacity'. Even 'spare capacity' is used, although this can be misleading because the capacity is not spare in the sense of being redundant and will certainly need replacing when it reaches the end of its life. Interpreting the meaning of 'backup capacity' in the literature or debate, therefore, requires an understanding of the definition to be used.

For any given demand, new conventional thermal plant capacity added can replace existing conventional plant capacity on a one-for-one basis; but obviously with a variable source such as wind, this equality relationship does not hold. In the last case shown in Figure 1.13, for example, although 25,000MW of wind capacity were added to the system, only 4500MW were retired and 20,500MW of conventional thermal plant were retained. These figures were calculated from simulations and based on a probabilistic risk assessment; but, as with all simulated results, they may or may not be valid.

Recent simulated power generation results for 25GW of wind generation across Britain (Renewable Energy Foundation, 2006) have been based on Renewable Obligation Certificate (ROC) data from the office of the electricity industry regulator, Ofgem, and correlated with historic wind data (Meteorological Office). The results indicate that over the period of 1995 to 2006, on average, wind power in January would have varied by 94 per cent of installed capacity, with power swings of 70 per cent of capacity over 30 hours being commonplace. On average, the minimum output would be only 3.7 per cent. Of more significance here are the maximum changes of national output of 99 per cent of capacity in 1998 and 1999, and the minimum outputs of 0.6 per cent of capacity in 1999 and 1 per cent of capacity in 2006. In such circumstances, what should be the level of conventional plant retained for *power* backup purposes – 100 per cent of the wind capacity or less?

Suffice it to say that for all practical purposes, there is a need for conventional backup capacity appropriate to the risks assumed regarding the acceptability of loss of supply of either power or energy. If the risk of loss of supply of *power* is captured, measured and effectively removed from further consideration by the use of a probabilistic 'capacity credit', as described above, whatever the practical shortcomings of such an analysis, then 'backup capacity' can be associated entirely with matching the potential loss of supply of *energy*. It is only in this latter connection that the requirements for backup capacity are explored further here; thus, in this chapter 'backup capacity' means that capacity required to ensure annual *energy* requirements are met. A similar restricted use of meaning is found elsewhere (UKERC, 2006)

Backup capacity and grid energy demand requirements

Conventional generation can be considered to provide two services: energy production and power capacity. If wind generation displaces conventional base-load plant power capacity by reference to capacity credit and probabilistic security of power supply standards, as described, then the provision of energy also has to be re-examined, and backup capacity to ensure energy supply must be provided if needed.

The relationship between security of supply, wind penetration, plant load factors and backup capacity may be mapped, as follows, using a simplified approach set out in Annex D of the System Costs of Additional Renewables (SCAR) report (ILEX Energy Consulting, 2002):

> If wind can provide no contribution to the required system capacity, then to be equivalent to the conventional generation, wind would require backup from generation equal to the conventional generation. This capacity could come from a number of sources, including old conventional generation or new open-cycle gas turbines (OCGTs). If, however, wind does contribute to system security, albeit at a lower rate than conventional capacity, then the above backup capacity requirement is reduced by the level of that contribution.

Suppose in the example shown in Figure 1.13 the 25GW of wind were added to the system with a market expectation of supplying energy over 8760 hours per year and operating, on average, according to an historic load factor – say, the national annual load factor for onshore wind power of 30 per cent:

$$\text{Load factor} = \frac{\text{(MWh generated pa)}}{\text{(MW nameplate capacity} \times 8760 \text{ hrs)}}. \qquad [4]$$

In Figure 1.13, the 25GW of wind power installed would be expected to yield $25 \times 8.76 \times 0.3 = 65.7$TWh to add to the total provided by the existing system.

The same annual generation of electrical energy could be provided by 8.8GW of conventional plant operating at an average 0.85 load factor.

If no capacity contribution is attributed to wind, then to ensure that the annual 65.7TWh would be delivered, the (energy) backup capacity equivalent to the conventional capacity would be 8.8GW. Such a circumstance would be akin to providing 100 per cent backup capacity (i.e. 100 per cent of the equivalent thermal plant capacity) and could arise in circumstances where the social and economic consequences of load exceeding supply are considered as the guiding rule, not the probabilities (LOLP) of such events.

Alternatively, if the 25GW of wind contributes 4.5GW of capacity to the system, then the additional backup capacity requirement is reduced by this amount and now becomes 8.8GW – 4.5GW = 4.3GW.

Assuming that any economically feasible existing generation would already be utilized on the system, then for the purposes of calculating standby plant costs, this extra capacity required would be, for example, OCGT plant (ILEX Energy Consulting, 2002).

The calculation can be summarized as follows:

$$
\begin{aligned}
&\text{X1 = wind capacity installed (GW) = 25} \\
&\text{X2 = wind capacity credit (GW) = 4.5} \\
&\text{X3 = wind load factor, LF(w), = 0.3} \qquad\qquad\qquad\qquad [5]\\
&\text{X4 = thermal plant load factor, LF(th), = 0.85} \\
&\text{X5 = thermal capacity equivalent of wind (GW)} \quad = \text{X1} \times \text{X3/X4} \\
&\qquad\qquad\qquad\qquad\qquad\qquad\qquad\qquad\qquad\qquad\quad = 25 \times 0.3/0.85 \\
&\qquad\qquad\qquad\qquad\qquad\qquad\qquad\qquad\qquad\qquad\quad = 8.8\text{GW} \\
&\text{X6 = required backup thermal capacity (GW)} \quad = \text{X5} - \text{X2} \\
&\qquad\qquad\qquad\qquad\qquad\qquad\qquad\qquad\qquad\qquad\quad = 8.8 - 4.5 \\
&\qquad\qquad\qquad\qquad\qquad\qquad\qquad\qquad\qquad\qquad\quad = 4.3\text{GW}.
\end{aligned}
$$

In this example, therefore, an additional 4.3GW of backup plant would be needed to guarantee delivery of *energy* over the year. The security of *power* supply standard has already been met by reference to the capacity credit, so this extra backup capacity would add further to the security of power supply.

The relationship shown between wind capacity installed and wind capacity credit, as shown in Figure 1.14, affords further insights into the related energy-supply backup capacity requirements.

Assuming that the wind capacity credit = $\sqrt{}$(GW of wind installed) as an approximation of the relationships shown in Figure 1.14, then the above calculation may be expressed as follows.

The required backup thermal capacity (Y) is the thermal capacity equivalent of the wind capacity minus the wind capacity credit, or:

$$\text{Y} = [\text{X1} \times \text{X3/X4}] - \sqrt{\text{X1}}. \qquad\qquad [6]$$

In the limit, when this required backup capacity is zero, Y = 0 and:

$$\text{X1} \times \text{X3/X4} = \sqrt{\text{X1}} \qquad\qquad [7]$$

or:

$$\text{X1} = [\text{X4/X3}]^2. \qquad\qquad [8]$$

The number of installed gigawatts of wind capacity when backup = 0 is represented by:

$$[\text{LF(th)/LF(w)}]^2. \qquad\qquad [9]$$

Figure 1.15 shows this last relationship between wind load factors and installed wind capacity for various thermal load factors. The curves denote combinations of wind capacity and wind load factors where no backup capacity is required. Of interest is the division of the space into areas where no backup capacity is needed below the curves, and vice-versa above the curves. Figure 1.16 shows the added relationships with annual wind energy generated at various load factors. By way of example, it is indicated that for an average

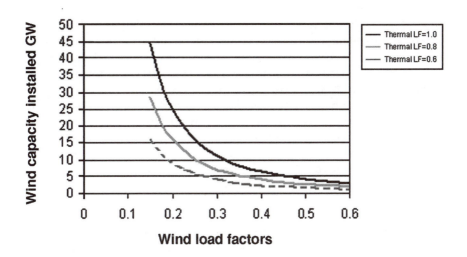

Figure 1.15 *Graphs of zero backup capacity separating the regions where extra backup capacity is not required (*left*) and backup capacity is required (*right*)*

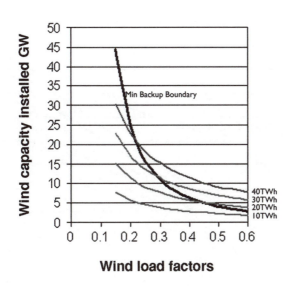

Figure 1.16 *Wind generation (TWh) and corresponding wind capacity for different load factors*

annual wind load factor of 30 per cent, any installed wind capacity above about 11GW will require extra backup capacity to be installed. The level of generation corresponding to the zero backup condition is about 30TWh, as seen in Figure 1.16, or approximately 7.5 per cent of national demand. The level of backup capacity and, hence, cost increases both with the wind capacity installed and with wind load factor for any given wind capacity.

The curve for a hypothetical thermal plant load factor of 1.0 represents a limiting condition above which it is not possible to operate without extra backup capacity. With the curves moving to the left with decreasing thermal plant load factor, and the backup capacity requirements increasing with the distance above the curves, this curve represents the boundary for the minimum backup capacity in the system, if any.

The backup capacities can be calculated now from Equation 6 for a range of possible installed wind capacities, all measured in gigawatts. From Equation 6 it can be seen that the ratio of the wind load factor to the thermal plant load factor is an important parameter determining the thermal plant equivalent capacity and, thus, the backup capacity. Figure 1.17 shows the different backup capacities needed for different installed wind capacities and different ratios of load factors from 0.25 to 0.375.

A feature of this graph is that for low wind capacities, here below 7GW, no backup capacity is required. Indeed, the addition of such wind capacity increases security of supply without detracting from the energy supply commitment. Above these levels, however, depending upon the wind and displaced thermal plant load factors, the backup capacity requirement increases with increasing installed wind capacity.

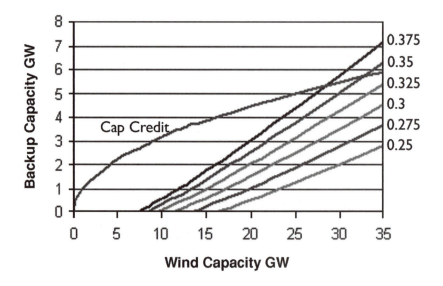

Figure 1.17 *Relationship between backup capacity and installed wind capacity for different ratios of wind load factor to thermal plant load factor*

The backup capacity needed can be seen to decrease as the ratios decrease (i.e. as the wind load factor decreases or the displaced thermal plant load factor increases). In some ways, this result may seem to be counterintuitive; however, it should be remembered that backup capacity is linked primarily to the supplying of energy, not to providing extra capacity for security of power supply. Wind capacity is added to the system with an expectation of energy supply determined according to the historic wind speeds reflected in the annual load factors. The lower the wind power load factor, the less generation is anticipated. As a result, the displaced thermal plant equivalent capacity is lower and, in turn, less backup capacity is required.

Shown also in Figure 1.17 is the capacity credit (i.e. the displaced thermal plant capacity), calculated using the approximate rule of the square root of the installed wind capacity in gigawatts. This depicts the general convergence of displaced thermal plant and added backup plant capacities with increasing levels of installed wind capacity.

CONCLUSIONS

Renewable energy forms an important component in future energy supplies for the electricity supply industry, the more so in the UK with increasing dependence upon imported gas and the retirement of coal and nuclear stations. Integrating renewable sources within the electricity supply system, however, requires special attention being paid to the specific variable power characteristics relevant to each respective source of energy and to the degree of penetration in meeting the system demand.

Inevitably, within the mix of renewable technologies in the future, a major role for wind generation is anticipated. The understanding of the nature and effects of the variability of wind power generation is therefore an important issue in ensuring both security of electrical power supply and security of electrical energy supply, two very different but linked problems.

The comprehensive studies referred to in this chapter examining the interaction of wind power with the UK grid system of supply produce some interesting results that may be summarized as follows:

- For the purposes of preserving security of power supply standards (defined in terms of loss of load probability, or LOLP) as wind capacity is installed in the system, the conventional planning margin capacity required is reduced by the capacity credit of wind generation.
- Detailed simulation and reliability analyses show that the capacity credit will never be more than the planning margin. This means that the total conventional plant capacity will never be less than the peak load irrespective of the amount of added wind capacity (a surprising result).
- As a consequence of the variable output, it is seen that wind power – and, by implication, the outputs of other renewable sources as well – can replace energy supplied from conventional sources, but not the need for most of their capacity. This will be a central problem for future studies and research.

- The immediate conclusion is that until new solutions emerge that will add substantially to the overall capacity credit of a more varied combination of variable energy sources, perhaps including very substantial energy storage capacities, much otherwise uneconomic conventional plant will need to be retained or replaced, either running on low or minimum output, or to be replaced by plant capable of frequent rapid start and ramping of output, such as (aero-derivative) OCGT generators (see also Chapter 6 in this volume).

These conclusions place an especially onerous security of supply requirement on market-driven investment and are perhaps not widely appreciated. They also represent a challenge to devising new methods for reducing the impact of the variability of renewable sources connected to the grid.

REFERENCES

Bell, K., Ault, G. and McDonald, J. (2006) 'All eyes on wind', *IET Power Engineer*, June/July, pp30–33

Boyle, G. (ed) (2004) *Renewable Energy*, 2nd Edition, OUP and the Open University, Oxford

E.ON Netz (2004) *E.ON Netz Wind Report: Wind Year 2003 – An Overview*, E.ON Netz Gmbh, Bayreuth, info@eon-netz.com

Grubb, M. J. (1986) *The Integration and Analysis of Intermittent Sources on Electricity Supply Systems*, PhD thesis, Cambridge University, Cambridge

Grubb, M. J. (1987) 'Capital effects at intermediate and higher penetrations', in *Proceedings of a Colloquium in the Electrical Engineering Department on Economic and Operational Assessment of Intermittent Generation Sources on Power Systems*, Imperial College, London, 5 March

Grubb, M. J. (1988) 'On capacity credits and wind-load correlations in Britain', in *Proceedings of the 10th BWEA Conference*, London, March

Halliday, J., Lipman, N., Bossanyi, E. A. and Musgrove, P. E. (1983) 'Studies of wind integration for the UK national grid', Wind Workshop VI, Minneapolis, US, June

ILEX Energy Consulting (2002) *Quantifying the System Costs of Additional Renewables in 2020*, a report to the Department of Trade and Industry (The 'SCAR' Report), Oxford, October

Laughton, M. A. (1990) (ed) 'Renewable energy sources', Watt Committee on Energy, Report no 22, Elsevier Applied Science, London

Laughton, M. A. (2002) 'Renewables and the UK Electricity Grid supply infrastructure', *Platts 'Power in Europe'*, no 383, 9 September, pp9–11

Meteorological Office (2006) Data describing wind speeds on a half-hourly basis for the UK, www.metoffice.com/education/archive/uk

National Grid (2002) *Supplementary Submission to the Performance and Innovation Unit (PIU) Report*, 28 September, National Grid, London

National Grid (2003) *Evidence presented to House of Lords Science and Technology Committee*, 10 December, National Grid, London

National Grid (2006) 'Seven year statement', www.nationalgrid.com/uk/Electricity/Data/

OXERA (Oxford Economic Research Associates) (2003) *The Non-Market Value of Generation Technologies*, Report for British Nuclear Fuels Limited (BNFL), OXERA, Oxford, 30 June

Palutikof, J. P. and Watkins, C. P. (1987) 'Some aspects of windspeed variability and their implications for the wind energy industry', in *Proceedings of a Colloquium in the Electrical Engineering Department on Economic and Operational Assessment of Intermittent Generation Sources on Power Systems*, Imperial College, London, 5 March

Renewable Energy Foundation (2006) Six files of data relating to the performance of the UK's renewables fleet; see also the seventh study (ref.wind.smoothing.08.12.06. pdf) reporting on the modelling of the behaviour of 25GW of wind, www.ref.org.uk/energydata.php

Rockingham, A. P. (1980), 'System economic theory for WECS', in *Proceedings of the 2nd BWEA Wind Energy Workshop*, Multi-Science, London

Royal Academy of Engineering (2002) *An Engineering Appraisal of the Policy and Innovation Unit's Energy Review*, Memorandum prepared for B. Wilson MP, Minister of State for Energy and Industry, London, August

Select Committee on the European Communities (1988) 'Alternative energy sources', 16th Report of the Select Committee on the European Communities, House of Lords paper 88, London, 21 June

Stones, J. (2003) 'Power quality', in Laughton, M. A. and Warne, D. F. (eds) *Electrical Engineers Reference Book,* 16th edition, Newnes, Oxford, Chapter 43

Swift-Hook, D. T. (1987) 'Firm power from the wind', in *Proceedings of the 9th BWEA Conference*, Edinburgh

Tyndall Centre (2003) 'Renewable energy in the UK', www.tyndall.ac.uk/publications/working_papers/wp22.pdf

UKERC (UK Energy Research Centre) (2006) *The Costs and Impacts of Intermittency*, UK Energy Research Centre, London, p112

Wetterzentrale (undated) Geographical coverage of anticyclones across northern Europe, including the UK, www.wetterzentrale.de/archive/2006/brack/bracka20061219.gif

2

Wind Power on the Grid

David Milborrow

INTRODUCTION

This chapter assesses the impact of wind on electricity networks, drawing upon information from around the world. It considers, first, how integrated electricity systems are managed, and notes that thermal power sources are not 100 per cent reliable. The characteristics of wind power are then discussed, and the way in which geographical diversity smoothes the output from wind farms is quantified.

It is shown that modest amounts of input from sources such as wind into a network pose no operational difficulties because they do not add significantly to the uncertainties in the prediction of the supply–demand balance. A review of integration studies, worldwide, suggests that the additional costs of integrating wind are around UK£2 per megawatt hour (MWh) with 10 per cent wind, rising to UK£3/MWh with 20 per cent wind. The role of storage, as a possible means of backup, is discussed.

The chapter also describes the consensus that wind plant has a 'capacity credit' and therefore can displace thermal plant. However, this credit declines – as a percentage of the wind capacity – as the penetration level increases, and this must be taken into account when evaluating the overall or 'total system' costs of assimilating wind. Estimates of these are made for a range of gas prices.

Most analyses of the impacts of wind energy have considered penetration levels below 20 per cent; but there is no reason why higher levels cannot be accommodated. The implications and costs of assimilating wind up to 100 per cent level are therefore examined.

Most of the supporting data is drawn from Britain and Denmark; but it is emphasized that most of the conclusions are also valid in the US and elsewhere. There is a brief discussion of factors that influence the costs of coping with variability and that affect the level of capacity credit.

ELECTRICITY SYSTEM OPERATION

'What happens when the wind stops blowing?' is a perennial question. It is often suggested that operating an electricity system with inputs from variable renewable sources, such as wind, is difficult. In a nutshell, that is incorrect. Nobody seems to ask: 'What happens when a nuclear power station suddenly shuts down?'

Numerous studies have examined the feasibility of operating power systems with the so-called 'intermittent' sources of renewable energy, usually wind, in the UK, Europe, Australia, the US and elsewhere. The first point to be made, however, is that wind is 'variable', not 'intermittent'. It is the output from 'conventional' sources of power that is intermittent. Although their characteristics vary considerably, problems with mechanical and electrical equipment, or with instrumentation, mean that sudden shutdowns – when up to 1000 megawatts (MW) or more of generation trips offline more or less instantaneously – are not uncommon. To illustrate the point, Figure 2.1 shows data from the cross-channel link between England and France. Between January and June 2005, there were five trips with 'cause unknown', and outage times, which cover both planned and unplanned maintenance, varied between zero (in February and March) and 14,000 minutes in April. An electricity utility can expect a thermal plant to be out of action for about 170 hours a year due to unforeseen circumstances, while planned maintenance accounts, in addition, for about 600 hours. The figures vary between plant types and location; but typical data is quoted by the Danish Energy Authority (2005).

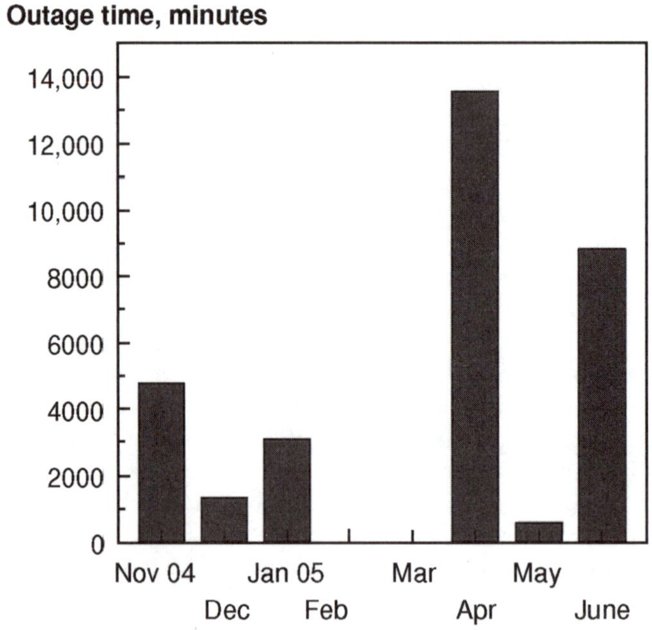

Outage time, minutes

Source: www.ucte.org

Figure 2.1 *Outages on the cross-Channel link (two 1000MW circuits) between England and France*

Figure 2.1 makes the point that there is no such thing as an electricity-generating plant that is 100 per cent reliable. Wind power output, in contrast to the output from a thermal plant, varies. Changes in wind speed never cause the power output from all the wind farms in a region or country to change by 100 per cent – up or down – instantaneously. Exactly how it does vary is discussed later. However, in order to appreciate the issues surrounding the integration of wind energy within an electricity network, it is necessary, first, to consider some key characteristics of integrated electricity networks. Next, the characteristics of wind energy systems are examined and, finally, the implications of operating electricity networks with wind.

The importance of aggregation

The effective operation of integrated electricity systems depends upon the aggregation of demand and generation. At one end of the spectrum, the minimum demand from a single house is a few watts; the average is about 0.5 kilowatts (kW) in the UK; and the maximum is 5kW to 10kW – 10 to 20 times the average. If each household met its own maximum demand (e.g. 5kW), 100 gigawatts (GW) of plant would be needed for this sector alone. Aggregation, however, smoothes variations in demand from all sectors; therefore, nationally, the maximum demand is around 60GW, about 1.5 times the average demand. As demands are added and smoothed, savings in generating plant are realized and load prediction becomes easier.

Aggregation can be illustrated using random number strings to simulate consumer demands. The 'demand' from one of ten consumers, together with the total, is illustrated in Figure 2.2. The single consumer's demand varies between 1kW and 9kW; but, when added together, ten consumers combine to produce fluctuations between 40kW and 70kW.

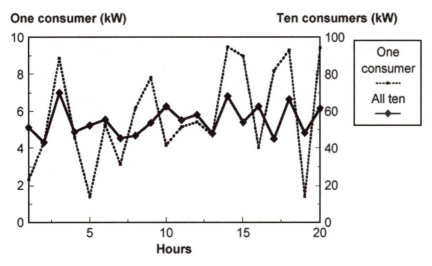

Figure 2.2 *Simple example of how aggregation promotes smoothing*

This discussion illustrates that the most efficient way of operating large electricity networks is to consider the integrated totality of the system. It does not make sense to consider any particular consumer demand in isolation or the output from any particular generation source. 'Levelling the output' of wind plants is often advocated; but this simply does not make sense if costs are to be minimized. To quote an American study (Hirst, 2001):

> A key feature of the present analysis [of the effects of variability] is its integration of wind with the overall electrical system. The uncontrollable, unpredictable and variable nature of wind output is not analysed in isolation. Rather, as is true for all loads and resources, the wind output is aggregated with all the other resources and loads to analyse the net effects of wind on the power system. Aggregation is a powerful mechanism used by the electricity industry to lower costs to all consumers. Such aggregation means that the system operator need not offset wind output on a megawatt-for-megawatt basis. Rather, all the operator need do, when unscheduled wind output appears on its system, is maintain its average reliability performance at the same level it would have been without the wind resource.

Managing uncertainty

Just as electricity networks must cope with unpredictable generating plants, they must also cope with unpredictable consumers. Although consumer demands from industry, commerce and the domestic sector can be forecast with reasonable accuracy, there is always the possibility of error. Television programmes may turn out to be more popular than expected, resulting in demand surges during the commercial breaks. Sudden changes in the weather can also cause unexpected surges – or drops – in demand.

The uncertainties in output from thermal plants and the uncertainties of consumer demand can be combined to produce a 'demand prediction error', or 'scheduling error'. The magnitude of the error depends upon how far ahead the prediction is made. The further into the future, the greater the error; but for one hour ahead, a typical scheduling error might be just over 1 per cent. Therefore, a system operator with a network similar in size to that managed by the California Independent System Operator might have a forecast demand for one hour ahead of, say, 25,000MW, ±300MW. That is the 'central estimate' of the error. It might be ±600MW, but with a lower probability, or ±900MW (three times the standard error), with an even lower probability. Most system operators schedule reserves in order to cover likely errors up to three or four times the standard error. So the operator would schedule 25,000MW of generation, plus about 900MW of reserve. These reserves are discussed later in this section.

The average daily errors in demand in a typical week on the English system are shown in Figure 2.3. During this period the maximum error in prediction

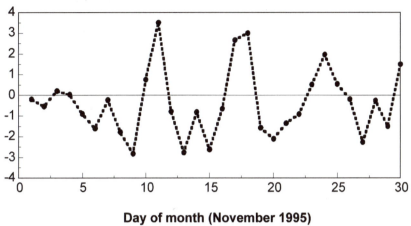

Day of month (November 1995)

Source: Electricity Pool

Figure 2.3 *Typical scheduling errors on the network in England and Wales*

was just under 4 per cent, and during 15 days it was less than 1 per cent. The standard error during this week was 1.6 per cent; since the average demand was about 32GW, this corresponds to about 500MW.

Power system defences

Large interconnected electricity systems have a number of robust defences against unexpected changes in the balance between demand and generation, including:

- *Inertia of the generating plant.* The mechanical and thermal inertia in the boilers and turbines of coal and nuclear power stations help to keep the power system stable. The contribution is small and passive, and is the first line of defence.
- *'Frequency response' plants.* These plants respond to frequency changes, automatically increasing or reducing output.
- *Reserve.* This refers to various types of plant. Some are operating at part load; some are off-line, but are able to start up within a short time:
  - *Pumped storage plant.* These can respond very rapidly to counteract any loss of generation or surge in demand. The UK power system, for example, has rapid response output from two such systems, with a maximum output of 2160MW.
  - *'Hot standby' plant.* This plant is able to provide generation on time scales ranging from a few minutes (in the case of gas turbines) to a few hours (in the case of a steam plant).

○ *'Standing reserve'*. This plant is available, if necessary, on a longer time scale and, as the name implies, is not necessarily operational all the time.
• *Voltage changes*. System voltage, like system frequency, is rarely 'spot on' its prescribed value, but varies within controlled limits. One response to a loss of generation, which may occur automatically or due to manual intervention, is a reduction of system voltage.

Voltage and frequency reductions can cope with demand or generation changes up to around 7.5 per cent, although only in exceptional circumstances would both be allowed to reach their minimum or maximum values.

The levels of reserve required at any given time depend partly upon uncertainties in the predictions of demand, but also upon the need to deal with the sudden loss of substantial amounts of generation, either due to power station faults or the loss of transmission circuits. In Britain, for example, key criteria are the possible loss of one circuit of the cross-Channel link (1000MW) or of Sizewell B nuclear power station (1200MW).

When scheduling reserves, system operators take into account the uncertainties in demand and generation on various time scales. Uncertainty increases with the time horizon; but, broadly speaking, the costs of the appropriate reserve decrease. Fast response plants may cost up to UK£5/MWh or more; but standing reserves may cost around UK£1/MWh.

WIND CHARACTERISTICS

Although the wind varies quite rapidly at any given location, wind turbines are excellent averaging devices since they react to the average wind over the whole of the blade circle. Groups of wind turbines in wind farms average the wind fluctuations over the whole of the wind farm; but network operators 'see' wind power fluctuations that are averaged over an even larger area. The greater the distances involved, the greater the smoothing as the correlation between wind speeds from different sites decreases with distance.

While the loss of the largest thermal unit on a power system (1200MW in the UK, but 400MW in Denmark) is a real risk, it is inconceivable that 400MW to 1000MW of dispersed wind generation will disappear instantaneously due to wind variations. The more wind-generating capacity that is installed, the more widely it is spread, and sudden changes of wind output across the whole country simply do not occur.

The way in which increased geographical spread reduces wind fluctuations is illustrated in Figure 2.4, which compares the output from a single 1000MW wind farm over a period of 24 hours with the output from 1000MW of wind distributed over England and Wales. Of necessity, this is based on a simulation (Holt et al, 1990).

A more rigorous way of presenting the data on fluctuation levels is shown in Figure 2.5 (Milborrow, 2001). This compares the percentage of time taken up by power changes within one hour recorded in western Denmark over a typical year, with the corresponding power changes recorded from a 5MW wind farm.

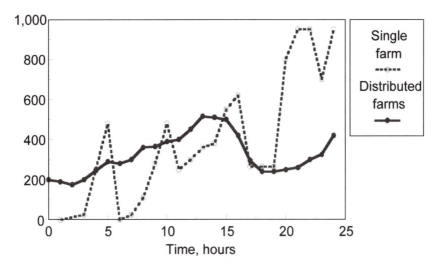

Source: Holt et al (1990)

Figure 2.4 *The smoothing effects of geographical dispersion (the 'single farm' and the 'distributed farms' both have 1000MW of capacity)*

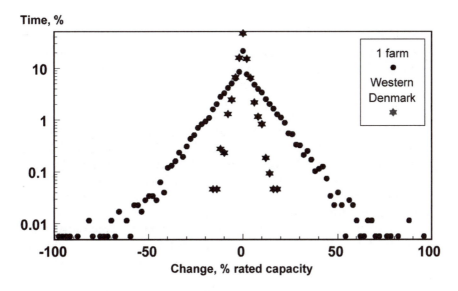

Source: Milborrow (2001)

Figure 2.5 *Comparison between the changes in output within an hour measured on a single wind farm and over the whole of western Denmark, in the first quarter of 2001*

Figures 2.4 and 2.5 illustrate the dramatic impact of geographical diversity. The data from western Denmark in Figure 2.5 shows that, for 78 per cent of the time, the power changes within one hour by less than ±3 per cent of its initial value. At the other end of the scale, the output from a single wind farm may, very occasionally, change by 100 per cent within an hour. In western Denmark, on the other hand, there were never any changes greater than 18 per cent. A very similar pattern of fluctuations has been observed in Germany (ISET, 2005).

This discussion has focused on power changes within an hour since these tend to have the strongest influence on the 'costs of variability'. System operators also take into account the additional uncertainty on longer time scales.

In the UK, the system operator National Grid Transco (NGT) has summarized the key issues relating to 'smoothing' as follows (National Grid Transco, 2004):

> However, based on recent analysis of the incidence and variation of wind speed, we have found that the expected intermittency of wind does not pose such a major problem for stability and we are confident that this can be adequately managed... It is a property of the interconnected transmission system that individual and local independent fluctuations in output are diversified and averaged out across the system.

NGT also discussed the implications of assimilating wind for the need for more reserve (National Grid Transco, 2004):

> Moreover, we do, and will continue to, carry frequency response such that frequency is contained within statutory limits for specified load and generation events... The interconnected transmission system enables this to be carried out more economically than would otherwise be the case.

> We believe that current levels of frequency response are sufficient even if the government's 2010 goal of 10 per cent of electricity supplies sourced from renewable fuels were all to be met by, say, wind technologies. In any event, should more response and reserve services be required, then our ancillary service market arrangements should encourage their cost-effective provision. We do not, therefore, foresee any significant technical problems arising from accommodating the government's targets for renewables and CHP [combined heat and power] by 2010.

MANAGING A NETWORK WITH WIND

The National Grid Transco quotation summarizes many of the key issues associated with assimilating wind on a network.

Modest amounts of wind cause few problems (or costs) for system operators since the extra uncertainty imposed on a system operator by wind energy is not equal to the uncertainty of the wind generation, but to the combined uncertainty of wind, demand and thermal generation. This combined uncertainty is determined from a 'sum of squares' calculation:

$$\sigma^2 \text{ (total)} = \sigma^2 \text{ (demand/generation)} + \sigma^2 \text{ (wind)}. \qquad [1]$$

To illustrate the point, the requirement for reserve in England and Wales at the winter peak is about 3500MW, based on uncertainties in demand and generation four hours ahead (Ofgem, 2004). Since system operators tend to schedule reserve by taking into account 'worst-case' scenarios, this figure is likely to be based on three times the standard deviation of the uncertainty (i.e. around 1200MW). The corresponding standard deviation in the uncertainty of wind generation, four hours ahead, is around 6 per cent; therefore, when Britain has 5000MW of wind generation, the standard error, four hours ahead, will be around 300MW. It follows that the *additional* standard error at periods of peak demand – using the sum of squares calculation in Equation 1 – will be around 37MW. It will clearly be higher at times of lower demand, but still modest.

Extra reserve and costs

The characteristics of most large electricity systems tend to be similar; therefore, estimates of the extra reserve needed to cope with wind energy are also similar. With wind supplying 10 per cent of the electricity, estimates of the additional reserve capacity are in the range 3 to 6 per cent of the rated capacity of wind plant. With 20 per cent wind, the range is approximately 4 to 8 per cent. Estimates of the 'extra costs of intermittency' are mostly close to National Grid's figures: accommodating 10 per cent wind on the UK system would increase balancing costs by UK£40 million per annum (UK£2/MWh of wind generation), and 20 per cent of wind generation would increase those costs by around UK£200 million per annum (UK£3/MWh of wind generation) (National Grid Transco, 2004). This data is shown in Figure 2.6, together with data from another UK study by ILEX Energy Consulting Ltd (2002) and four American studies (Coatney, 2003; Electrotek Concepts, 2003; Seck, 2003; EnerNex Corporation, 2004).

Since the data in Figure 2.6 is drawn from four American studies, the costs are quoted in US$/MWh of wind generation. Although there is some scatter, it may be noted that the additional cost is in the range US$1.6/MWh to US$3.5/MWh with 5 per cent wind energy, rising to between US$2.6 /MWh and US$4.6/MWh with 10 per cent wind energy. Since the generation cost of wind energy is now around US$60/MWh, the extra costs of variability at the 10 per cent level add less than 10 per cent to the overall generation cost.

Storage

When it comes to sourcing the most economic method of providing reserve, system operators choose the least cost options, provided that they meet their technical requirements. Storage has no intrinsic merits for coupling with wind

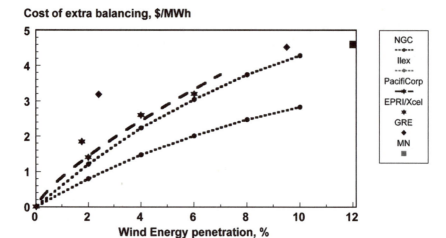

Figure 2.6 *Estimates of the cost of extra balancing needed for wind. Significant increases in fossil fuel prices from 2006 to 2008 have resulted in higher prices for reserves that use coal and gas plants. As a result, some of the latest estimates of extra balancing costs are at or above the top end of the range shown here*

energy, as an early analysis by Farmer et al (1980) made clear: 'there is no operational necessity in associating storage plant with wind-power generation, up to a wind output capacity of at least 20 per cent of system peak demand'.

This quote implicitly deals with the idea that storage might help to 'level the output' from intermittent renewables. This is possible; but it simply adds to wind's costs – unless the added value exceeds the extra cost. Storage may or may not be the most effective way of providing additional spinning reserve for the system – this, again, depends upon its costs.

The breakeven cost for storage is controlled by the pattern of electricity prices, the lifetime of the device and the rate of return required by the owner. The higher the difference between peak and off-peak electricity prices, the more one can afford to pay for storage. A UK study has suggested that the breakeven cost is in the range UK£500/kW to UK£800/kW – possibly higher if the technology can provide a range of services (Strbac and Black, 2004).

There was considerable interest in a new reversible 'flow battery' concept, pioneered by Innogy. This, it was claimed, had the potential to bring about a revolution in the power industry (Milborrow, 2000). Regenesys, as it was termed, was a reversible fuel cell, and the first commercial plant was to have a power output of 15MW and a storage capacity of 120MWh. Innogy expected the cost of its first commercial prototype to be around UK£1000/kW, somewhat higher than the 'target' cost quoted in the previous paragraph. Although work on the project has now halted, the technology has recently been sold. The concept is attractive since it holds out the prospect of 'utility-scale' storage. In most of the developed world, new sites for pumped storage are difficult to find.

Although there are some prospects for compressed air storage, this has yet to become established to any significant degree.

Demand-side management

Demand-side management has a similar role as storage: it may be the most economical way for system operators to provide reserve. It is an area of increasing interest, and ideas for remote control of non-essential consumer loads are being investigated. As with storage, there may be opportunities for links with wind energy developments, depending upon the economics. It may be a viable way of increasing the amount of wind generation that can be accepted onto a weak network (Econnect, 1996).

CAPACITY CREDITS

An issue that affects economic appraisals of variable renewable sources is the concept of 'capacity credit'. The capacity credit of any power plant may be defined as a measure of the ability of the plant to contribute to the peak demands of a power system. Capacity credit here is defined as the ratio of capacity of thermal plant displaced to rated output of wind plant.

Numerous studies of the UK network have concluded that wind plant has a capacity credit. An early study, already cited (Farmer et al, 1980), concluded:

> If a definition of capacity credit is adopted that maintains the existing level of security of supply, it can be shown that for low levels of wind-power penetration, a substantial proportion of the output can be ascribed as firm power... even at higher levels of penetration, the capacity credit could approach 20 per cent of the rated output.

More recently, the UK system operator has estimated that 8000MW of wind might displace about 3000MW of conventional (gas-fired) plant, and 26,000MW of wind (20 per cent penetration) would displace about 5000MW of such plant.

It should be noted that, strictly speaking, the reference plant should be specified. Thermal plant is not 100 per cent reliable, and so the 'firm power' equivalent of such plant depends upon the technology and may vary between utilities. If 1000MW of wind can provide, say, 340MW of firm capacity, this is equivalent to 400MW of combined-cycle gas turbine (CCGT) plant if the peak period availability of that plant is 0.85. If the reference plant was coal, with lower peak availability, the capacity credit might be higher.

Power system operations depend upon assessments of risk, and this plays an important role in the evaluation of capacity credit. No system is risk free; but plant is scheduled to keep these risks within defined boundaries. The most rigorous method of assessing the capacity credit of wind is to look at a power system with a known loss of load expectation (LOLE), which is the integrated value of the loss of load probability (LOLP) over a year, and add a time series that is representative of typical wind power output. This reduces the LOLE so that 'firm power' can be subtracted to restore it to its original value. Firm power divided by peak time availability of conventional plant equals the capacity credit. In practice, LOLE is heavily influenced by the loss of load probability at the time of system peak demand; therefore, acceptable accuracy in calculating capacity credit is often obtained simply by looking at the availability of wind power at this time – preferably over a period of many years. If wind and demand are not correlated, the average power output at any time (not just the winter peak) will be proportional to the capacity factor.

Theoretical treatment

For first-order accuracy, a simple mathematical analysis shows that the contribution of any item of power plant to firm capacity is equal to the average power that it can generate (Swift-Hook, 1987). This conclusion is only valid if there is no correlation between wind and electricity demand. Dale et al (2004) discussed this latter point and concluded that the assumption is reasonable. In most of northern Europe, the average capacity factor realized during the winter quarter, when peak demands occur, is slightly higher than the year-round average. As a result, values of capacity credit reflect this.

Several studies of the impacts of wind have addressed the issue in more detail, and their conclusions are succinctly summarized in a study carried out for the European Commission (Holt et al, 1990): 'At low [energy] penetration the firm power that can be assigned to wind energy will vary in direct proportion with the expected output at [the] time of system risk.' In practice, this statement is true for any energy source, whether it is renewable or not. 'Firm power' is not the same as 'capacity credit', as noted earlier; but the two are readily linked.

Estimates of the decline of capacity credit with increased levels of the intermittent renewable sources need further information about the characteristics of those sources and the electricity systems to which they are connected (Rockingham, 1979).

Effect of calms

It is often suggested that large high-pressure weather systems may prevent wind generation over the whole of a country during times of peak demand. This, it is argued, means that substantial amounts of backup are required to replace the 'missing' wind generation.

Focusing on Great Britain, there are two weaknesses in this hypothesis:

1   There does not appear to be any evidence that the whole country is regularly becalmed at times of peak demand. On the contrary, work carried out at the University of East Anglia (Palutikof et al, 1990) suggests that 'peak demand times occur when cold weather is compounded by a wind chill factor, and low temperature alone appears insufficient to produce the highest demand of

the year'. Other work carried out at Oxford University has reached similar conclusions (Sinden, 2005; see also Chapter 3 in this volume).

2   System operators with wind on their networks never rely on the full rated output of wind plant being available. Its expected contribution to peak demand is quantified by its 'capacity credit', which is the amount of thermal plant that can be displaced, leaving the power system with the same reliability. Roughly speaking, 1000MW of wind in the UK has a capacity credit of about 400MW. On average, roughly that amount of wind will be available at times of peak demand. On some occasions it will be less; on others will be more. This is no different from the approach to the capacity credit of, say, a nuclear plant. On average, the availability of this plant at times of peak demand is around 85 per cent of its rated output. On some occasions, however, it will be less; on others, more.

Results of analysis

Numerous studies have shown that, statistically, wind *can* be expected to contribute to peak demands. The evidence is very robust and there are several approaches:

*   The statistical approach, as formulated by Swift-Hook (1987) and others, assumes that winds are random in nature with respect to electricity demand, a point discussed above.
*   Analyses of wind turbine output: the work at the University of East Anglia, cited above, is very relevant. Using just four sites, they showed that the summed average wind turbine outputs during eight winter peak-demand periods were about 32 per cent of rated output. National Wind Power has similarly found that 'wind farm capacity factors during periods of peak demand are typically 50 per cent higher than average all-year capacity factors' (Warren et al, 1995).
*   Power system simulations include those of the Central Electricity Generating Board (CEGB) (Gardner and Thorpe, 1983; Holt et al, 1990, which used 12 meteorological sites), as well as more recent work by National Grid Transco (National Grid, 2001).

The UK studies have all yielded similar results, pointing to wind having a capacity credit roughly equal to the 'winter quarter' capacity factor, (around 30 to 40 per cent) at low penetrations. Thereafter, it decreases, reaching around 20 per cent, with 20 per cent wind on the network. Increased use of offshore wind, with higher wind speeds and greater geographical diversity, is likely to increase the firm power contribution.

Although similar results have come from other studies elsewhere, the findings are not universal, and higher or lower values of capacity credit are reported as they depend upon wind strengths and the correlation between wind and demand. The capacity credit of wind generation in most of northern Europe is roughly equal to the capacity factor in the winter quarter (Milborrow, 1996). Results from ten European studies are compared in Figure 2.7,

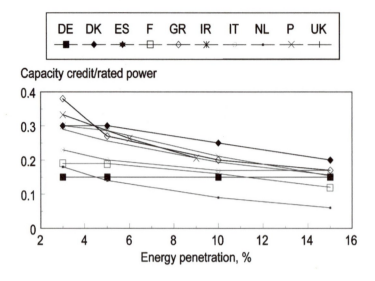

Figure 2.7 *Comparison of results from ten utility studies of capacity credit (note that some made arbitrary assumptions)*

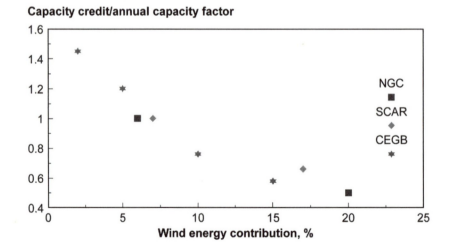

Note: The data labelled System Costs of Additional Renewables (SCAR) is derived from ILEX Energy Consulting Ltd (2002).

Figure 2.8 *Normalized values of capacity credit from three studies of the National Grid Company (NGC) system*

showing credits declining from 20 to 40 per cent at low wind penetrations, to 10 to 20 per cent with 15 per cent wind.

In order to facilitate comparisons between UK studies, Figure 2.8 compares normalized values of capacity credit and shows a good measure of agreement. With modest contributions of wind energy, the capacity credit is about 40 per cent greater than the annual capacity factor; therefore, if the average capacity factor across the country was 35 per cent, then 1000MW of wind would displace 490MW of thermal plant. At higher wind energy penetrations, the capacity credit declines so that, with 20 per cent wind, the credit is about half the capacity factor.

Wind-induced demands: The cooling power of the wind

It is known that a rise in wind speed may be accompanied by rising demand, and one estimate for the link between heating load and wind velocity suggests a linear relationship (Lacey, 1951). If so, this would enhance the value of wind energy. However, predictions of increased heating loads due to wind are empirical in nature. The difficulty of establishing precise estimates for rates of air leakage, together with the variations between different types of building, means that it is not practical to devise rigorous theoretical estimates; but the form of the equation used by the UK electricity industry suggests a square root relationship. This is broadly consistent with what might be expected from an examination of the underlying physical processes. Although some authors have failed to find evidence of a correlation, this does not necessarily imply that the phenomenon does not exist. One recent study appears to indicate that some correlations can be detected at the local level (South Western Electricity plc, 1994).

TOTAL EXTRA COSTS OF WIND ENERGY

The total extra costs of operating an electricity network with increasing amounts of wind are now of considerable interest due to the recent increases in the price of gas. The differences between the generation costs from wind and those from gas are narrowing, and it is quite likely that wind energy might become cheaper. The total extra costs of wind energy take into account the following:

- extra costs associated with variability, as discussed in the previous sections;
- extra generation costs (if any);
- extra costs of transmission and distribution.

During the last two or three years, several analyses have appeared that quantify the additional costs (if any) to electricity consumers of increasing the amount of renewables – especially wind energy – in the generation mix. Examples include an analysis for the UK (Dale et al, 2004), and for Pennsylvania (Black and Veatch Corporation, 2004). The latter suggested that a 10 per cent renewables portfolio by 2015 would increase costs by US$0.4/MWh. Wind contributed about 65 per cent of the renewables mix. The UK analysis suggested that the extra cost to the electricity consumer of providing 20 per cent of supplies from wind energy would be about UK£3/MWh. In the light of recent gas price rises, this estimate is now very pessimistic: as the price of gas goes up, so the 'fuel saving value' of wind energy also goes up. Dale et al (2004) used a UK-delivered gas price of about 19 pence per therm (p/therm). It is now difficult to quote future gas prices with any certainty, and a question mark hangs over estimates of future wind plant costs due to increases in wind turbine prices during 2005 and 2006. To deal with this uncertainty, Figure 2.9 shows breakeven installed costs for a range of gas prices, as well as wind energy penetration levels of 5, 7.5 and 10 per cent.

It should be noted that these breakeven prices assume that there are equal amounts of onshore and offshore wind energy. Figure 2.9 shows, for example, that for an onshore installed cost of UK£800/kW – roughly the average level in 2005 – the breakeven gas price is about UK£0.5/therm. During 2005, the average gas price was just under UK£0.3/therm, and the corresponding onshore wind cost (at the breakeven point) was just under UK£600/kW (offshore UK£900/kW). However, if the development of offshore wind continues to be slow and its contribution is negligible by the time a 5 per cent contribution is achieved, the breakeven gas price is UK£0.3/therm – the average UK price in 2005.

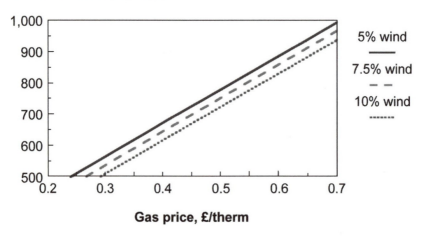

Onshore wind cost, £/kW

Gas price, £/therm

Note: Offshore/onshore split: equal; offshore cost 50% higher.
35% capacity factors.
Carbon dioxide: €24/tonne (Oct 2005).

Figure 2.9 *Breakeven wind costs for a range of gas prices*

Wind predictability

In general, wind is perceived as unpredictable; but this is not strictly true. Weather forecasting techniques enable average wind speeds to be predicted with reasonable accuracy, but not necessarily the hour-by-hour variations. In some parts of the world, however, such as the Californian passes, the winds do have an established diurnal cycle, which benefits system operators. Although there are significant differences in the way in which electricity jurisdictions operate, there is a reasonable consensus on the savings that can be realized through good forecasting. These savings accrue since the uncertainties that system operators face when handling wind energy are significantly reduced, and this enables them to lower the amount of extra reserve plant that is scheduled. Although the monetary savings depend upon the costs of reserve, a recent American study (Milligan, 2003) suggested that the requirements for backup reserve capacity, when the wind capacity amounts to 22.6 per cent of peak demand, might be reduced, through forecasting, from 7.6 per cent of the wind capacity to 2.6 per cent. This suggests that the costs of extra balancing might be lowered by a factor of about three, although a realistic target may be a halving of capacity requirements and costs. There is a considerable amount of worldwide activity aimed at improving forecasting accuracy for wind plant (Kariniotakis et al, 2003; see also Chapter 5 in this volume).

WIND ENERGY PENETRATION LEVELS ABOVE 20 PER CENT

Although the British and Danish system operators have both stated that the limits to wind energy penetration are economic rather than technical, this has gone largely unnoticed. There is still a feeling that high wind energy penetrations will cause severe technical problems; but this is simply not the case. Extra costs are incurred; but these can be quantified. A recent Danish study (Pedersen et al, 2006) has suggested that these extra costs reach a maximum value of around €15/MWh of wind. If markets can be found for the surplus wind (when wind output exceeds consumer demands), then that figure comes down.

The overall message is very clear. High wind energy penetrations can be accommodated on electricity networks without any 'step changes' in additional costs being incurred. The results from the Danish analysis (Pedersen et al, 2006) have been set alongside an analysis for the UK system using the methods of Dale et al (2004). The results from the two studies show a good measure of agreement. Even with 60 per cent wind, the 'variability premium' may be completely offset by cheaper wind energy generation costs (Milborrow, 2006).

As the wind energy penetration rises above 20 per cent, there will be occasions when the wind power output exceeds the electricity demand. Various options are then possible. It may be possible to sell the power to a neighbouring utility or, at some time in the future, use it to generate hydrogen for road transport. Alternatively, it may make sense to charge a storage facility, whether that is pumped storage, a flow battery or a compressed air store. However, the storage option demands a critical look at the economics. The storage facility must be paid for, and so the electricity sent out from the store has a higher cost than the electricity used to charge it. The efficiency of the storage device must also be taken into account. If the higher cost of the electricity leaving the store is still economically attractive to the electricity network, then storage can be used. If not, and if no other option is available, then the output from wind turbines may need to be curtailed.

The point at which surplus wind power may need to be rejected or diverted to other markets will depend upon the capacity factor of the wind and the electricity demand pattern. This situation is being observed, very occasionally, in western Denmark now that the penetration level has reached 23 per cent. The surplus wind energy increases from 0.5 per cent with 30 per cent wind, to 3.5 per cent with 50 per cent wind, 17.5 per cent with 80 per cent wind, and 30 per cent with 100 per cent wind. Because of the surplus wind energy, the nominal '100 per cent wind' case actually delivers only 70 per cent wind to consumers. It would, in theory, be possible to increase the wind capacity even further; but the economics become progressively less attractive and so this is not regarded as realistic.

Despite the existence of the surplus, the total carbon dioxide (CO_2) emissions from electricity generation fall steadily as more wind is added. By the time the (nominal) 100 per cent wind penetration level is reached, the emissions are around 30 per cent of the 'no wind' level.

Counting the cost at high penetrations

The key question to be addressed is the cost of variability. This is simply the cost of the extra balancing, plus the extra generating costs of the thermal plant due to its reduced load factor. The results from the Danish study may be compared with an analysis using the UK methodology of Dale et al (2004). The UK analysis assumed that 26,000MW of wind (20 per cent penetration) would have a capacity credit of 5000MW. The Energinet study (Pedersen et al, 2006), by contrast, assumed a significantly lower capacity credit – about 300MW maximum – and this accounts for most of the difference between the final results, shown in Figure 2.10. With 10 per cent wind penetration, the UK methodology predicts a 'cost of variability' of €4/MWh of wind, compared with Energinet's figure of €10/MWh. The disparity narrows at higher wind energy penetrations. With 50 per cent wind, the UK analysis suggests €11/ MWh, which compares with €14.5/ MWh from the Danish analysis. With 100 per cent wind, the figures are €13.3/MWh and €14.8/MWh, respectively.

Possible roles for storage and hydrogen

There are three possible options for dealing with the wind energy that is surplus to requirements. The wind turbines may simply have their output curtailed or, if the electricity system has links with other networks, it may be

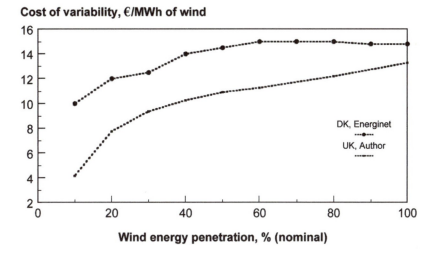

Figure 2.10 *Costs of variability (per unit of wind energy) at high wind penetrations*

possible to export it. This option is certainly possible in the case of western Denmark, although it may not be possible to export all the power at very high penetration levels if the magnitude of the surplus power exceeds the capacity of the links. The third option is to divert the surplus energy into new markets. The lower the price at which the surplus electricity is sold, the greater the chances of finding a market. Energinet assumed a modest value of €13.3/MWh for the surplus, which reduced the maximum cost of the variability penalty from €15/MWh to around €13.5/MWh (with 50 per cent wind), falling to €9/MWh with 100 per cent wind.

Energinet does not suggest using hydrogen as a storage medium with a view to feeding electricity back into the system; but it could be used for transport. That would necessitate additional plant, such as fuel cells, and incur significant losses; so it is unlikely that it would be economic.

If electricity markets with high wind penetration do become established, the prospects for storage technologies, such as flow batteries and compressed air, may improve. It is doubtful whether it would often make economic sense to 'level the output' of wind power simply because the additional value is unlikely to exceed the additional cost. On the other hand, a storage system that can be charged with electricity at below market cost can possibly be used for arbitrage – releasing electricity into the system when market prices are high.

THE INFLUENCE OF NATIONAL AND REGIONAL DIFFERENCES

The effects of wind speed and scheduling procedures

The characteristics of wind tend to be similar the world over and the same applies to electricity networks – but there are some differences. The extra costs of reserve in Great Britain and a number of American electricity jurisdictions are similar (see Figure 2.6); but significantly higher figures are quoted for Germany. This is due partly to the way in which the electricity network is operated, and partly to the lower wind speeds that prevail there. In Germany, as well as in some other electricity jurisdictions, wind tends to be treated in the same way as gas or coal-fired plant and is required to forecast its output several hours ahead. If the plant schedules lack flexibility, it is quite likely that the output from the wind plant will change after the commitment is made. This may mean that balancing power must be purchased to make up any power deficits or, alternatively, surplus wind may need to be sold for a low price. The more flexibility that is built into plant scheduling, the more efficiently the system can be operated.

The other reason why balancing costs for wind in Germany tend to be higher than elsewhere is that wind speeds are quite low. The average capacity factor of German wind generation is about 15 per cent, which compares with 25 per cent in Denmark and 30 per cent in Britain. This means that the wind

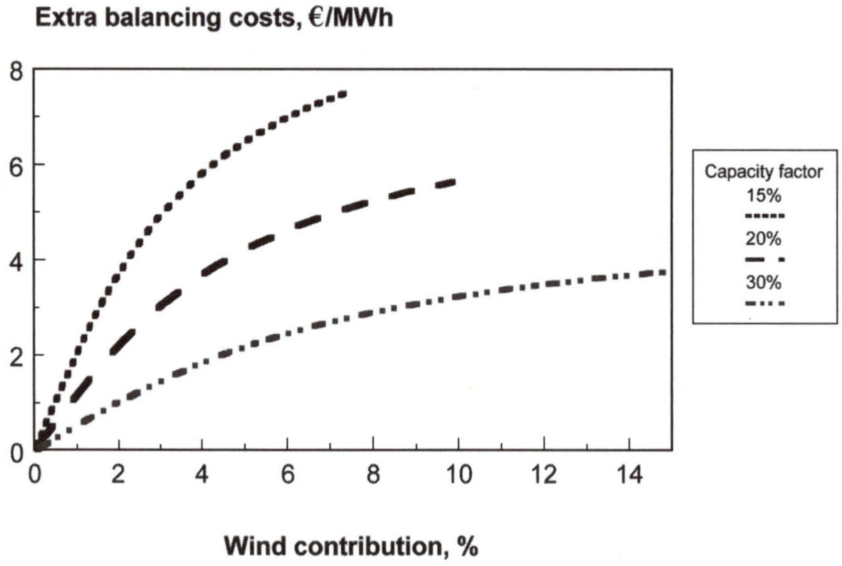

Extra balancing costs, €/MWh

Wind contribution, %

Figure 2.11 *The influence of capacity factor on extra reserve costs*

plant capacity needed to generate a given amount of energy in Germany is roughly twice the capacity needed in Britain. The magnitude of the power swings from the plant in Germany will therefore be higher than those in Britain, and the additional uncertainty means that the system operator needs to schedule more reserve. The impact of capacity factor on extra reserve costs is illustrated in Figure 2.11. This shows that the additional costs associated with 5 per cent wind energy penetration are around €2/MWh in a region where the capacity factor is 30 per cent, but nearly €6/MWh in a region where the capacity factor is 15 per cent. A very thorough analysis of the impacts of wind on the German network was completed in 2005 (DENA, 2005).

Predictable winds

While predictable diurnal variations may reduce the costs of additional reserve, as noted earlier, they do not necessarily enhance the overall value of wind energy. That is because peak wind power may systematically occur at times that are not consistent with peak consumer demands. On the other hand, if the reverse is true, this increases the capacity credit of wind plant, and one Californian study suggested that the capacity credit of one of the early machines was about 80 per cent of its rated power (Pacific Gas and Electric Company, 1989). At the other end of the spectrum, a study for the North West Power Pool suggested a value for that region of only 5 per cent (Flaim and Hock, 1984).

Transmission issues

One factor that tends to be very specific to particular utilities comprises the transmission constraint and reinforcement costs. Not only do connections need to be provided for new wind plant, but there is also the possibility that extra reinforcements may be needed to cope with transmission congestion. This occurs when significant amounts of wind are concentrated in the windiest regions, with the result that there is insufficient capacity to transmit all the wind power when it is operating at or near peak output. These issues can only be studied on a case-by-case basis.

CONCLUSIONS

Although the impacts of wind variability are sometimes controversial, it is doubtful whether any compelling evidence has been provided to invalidate the substantial amount of work that has been carried out on this topic. This chapter only draws on a fraction of the references that were cited in a very thorough literature survey completed recently (Gross et al, 2006; see also Chapter 4 in this volume), and more information has been published since the report was completed. The conclusions are unchanged: wind variability can be managed, technically and at modest cost. The cost amounts to around UK£2/MWh for 10 per cent wind and may fall as better wind forecasting techniques emerge.

This is only a small fraction (about 5 to 10 per cent) of the generation cost of wind energy, so that the impact of changes in fuel price on backup costs are unlikely to have a substantial impact on the cost-competitiveness of wind energy.

The recent increases in the price of gas have made a substantial difference to the 'total extra cost' of operating the power systems with wind energy. Although future gas price trends are very difficult to predict, it is quite likely that the overall cost to electricity consumers in systems with substantial amounts of wind energy will be small and possibly negative – in other words, savings will be made due to the lower generating costs of wind energy.

As more experience is gained in operating electricity networks with wind energy, it is likely that the additional costs associated with variability will fall. In addition, improved techniques of demand-site management are likely to reduce the costs of managing electricity networks and, as a consequence, the additional costs of managing variability. Last, but not least, worldwide research into improved methods of wind predictability are also likely to reduce the costs of integrating wind energy.

REFERENCES

Black and Veatch Corporation (2004) *Economic Impact of Renewable Energy in Pennsylvania*, Black and Veatch Corporation, Kansas City, Kansas

Coatney, T. (2003) 'Modeling wind energy integration costs', UWIG Technical Wind Workshop, Seattle, WA, www.uwig.org/TechnicalWorkshop03-wa.html

Dale, L., Milborrow, D., Slark, R. and Strbac, G. (2004) 'Total cost estimates for large-scale wind scenarios in UK', *Energy Policy*, vol 32, pp1949–1956 (first published as Dale, L., Milborrow, D., Slark, R. and Strbac, G. (2003) 'A shift to wind is not unfeasible', *Power UK*, no 109, pp17–25)

Danish Energy Authority (2005) 'Technology data for electricity and heat generating plants', Danish Energy Agency, www.ens.dk

DENA (Deutsche Energie-Agentur) (2005) *Planning of the Grid Integration of Wind Energy in Germany Onshore and Offshore up to the Year 2020*, Deutsche Energie-Agentur, Berlin

Econnect Ltd (1996) *Wind Turbines and Load Management on Weak Networks*, ETSU W/33/00421/REP, Future Energy Solutions, Harwell

Electrotek Concepts (2003) 'Characterising the impacts of significant wind generation facilities on bulk power system operations planning – EPRI/Xcel: A 'case study' for Xcel Energy', www.uwig.org/TechnicalWorkshop03-wa.html

EnerNex Corporation (2004) *Wind Integration Study for Xcel Energy and the Minnesota Department of Commerce*, EnerNex Corporation, Minnesota

Farmer, E. D., Newman, V. G. and Ashmole, P. H. (1980) 'Economic and operational implications of a complex of wind-driven power generators on a power system', *IEE Proc A*, vol 127, no 5, pp289–295

Flaim, T. and Hock, S. (1984) *Wind Energy Systems for Electric Utilities: A Synthesis of Value Studies*, Solar Energy Research Institute, SERI/TR-211-2318, Solar Energy Research Institute, Boulder, CO

Gardner, G. E. and Thorpe, A. (1983) *System Integration of Wind Power Generation in Great Britain*, EC Contractors' Meeting, Brussels, 23–24 November 1982, D.

Reidel Publishing, Lancaster, UK

Gross, R. Heptonstall, P., Anderson, A., Green, T., Leach, J. and Skea, J. (2006) The *Costs and Impacts of Intermittency*, UK Energy Research Centre, London

Hirst, E. (2001) *Interactions of Wind Farms with Bulk-Power Operations and Markets*, Prepared for Sustainable FERC Energy Policy, Virginia

Holt, J. S., Milborrow, D. J. and Thorpe, A. (1990) *Assessment of the Impact of Wind Energy on the CEGB System*, CEC, Brussels

ILEX Energy Consulting (2002) *Quantifying the System Costs of Additional Renewables in 2020*, The 'SCAR' Report to the DTI, Oxford, October

ISET (Institut für Solare Energieversorgungstechnik) (2005) *Wind Energy Report*, ISET, Germany

Kariniotakis, G. et al (2003) 'ANEMOS: Development of a next generation wind power forecasting system for the large-scale integration of onshore and offshore wind farms', Paper presented at European Wind Energy Conference, Madrid

Lacy, R. E. (1951) 'Variations of the winter means of temperature, wind speed and sunshine, and their effect on the heating requirements of a house', *Meteorological Magazine*, vol 80, pp161–165

Milborrow, D. J. (1996) *Capacity Credits – Clarifying the Issues*, British Wind Energy Association, 18th Annual Conference, Exeter, 25–27 September, MEP Ltd, London

Milborrow, D. J. (2000) 'Revolutionary potential', *Windpower Monthly*, vol 16, p10

Milborrow, D. J. (2001) 'Penalties for intermittent sources of energy', Working paper for UK Energy Review, www.pm.gov.uk/files/pdf/Milborrow.pdf

Milborrow, D. J. (2006) 'The economic myths of high wind penetration', *Windpower Monthly*, September, vol 22, pp51–56

Milligan, M. (2003) *Wind Power Plants and System Operation in the Hourly Time Domain*, National Renewable Energy Laboratory NREL/CP-500-33955, Golden, CO

National Grid (2001) 'Energy policy review: Investigation of wind power intermittency', www.dti.gov.uk

National Grid Transco (2004) 'Submission to Scottish Parliament, Enterprise and Culture Committee, 'Renewable energy in Scotland' inquiry, www.scottish.parliament.uk/enterprise

Ofgem (2004) NGC system operator incentive scheme from April 2005: Initial proposals, www.ofgem.gov.uk

Pacific Gas and Electric Company (1989) *Solano MOD-2 Wind Turbine Operating Experience through 1988*, Electric Power Research Institute, GS-6567, Palo Alto, CA

Palutikof, J., Cook, H. and Davies, T. (1990) 'Effects of geographical dispersion on wind turbine performance in England: A simulation', *Atmospheric Environment*, vol 24A, p1

Pedersen, J., Eriksen, P. B. and Orths, A. (2006) 'Market impacts of large-scale system integration of wind power', Paper presented to European Wind Energy Association Conference, Athens

Rockingham, A. P. (1979) 'System economic theory for WECS', in *Proceedings of the 2nd British Wind Energy Association Conference*, Cranfield, Multi-Science, London

Seck, T. (2003) 'GRE: An analysis for great river energy', www.uwig.org/Technical Workshop03-wa.html

Sinden, G. (2005) *Wind Power and the UK Resource*, Environmental Change Unit, University of Oxford, Oxford, www.eci.ox.ac.uk/renewables

South Western Electricity plc (1994) *Interaction of Delabole Wind Farm and South Western Electricity's Distribution System*, ETSU Report W/33/00266/REP, Future

Energy Solutions, Harwell

Strbac, G. and Black, M. (2004) *Future Value of Storage in the UK*, Centre For Electrical Energy, Manchester

Swift-Hook, D. T. (1987) 'Firm power from the wind', in *Proceedings of the 1987 British Wind Energy Association Conference*, Edinburgh, 1987, Mechanical Engineering Publications Ltd, London

Warren, J., Hannah, P., Hoskin, R., Lindley, D. and Musgrove, P. (1995) 'Performance of wind farms in complex terrain', in *Proceedings of the 1995 British Wind Energy Association Conference*, Warwick, Mechanical Engineering Publications Ltd, London

3

Renewable Resource Characteristics and Network Integration

Graham Sinden

INTRODUCTION

The large-scale development of renewable electricity generation in the UK will require the integration of renewable electricity supplies within electricity networks, and robust analysis of the impact of such supplies is critical to evaluating the requirements and evolution of electricity networks into the future. Central to such analysis is an accurate understanding of the characteristics of renewable resources and, as a consequence, the characteristics of the electricity supply that they provide.

Key issues surrounding renewables include the seasonal distribution of the resource, availability during times of peak demand, the variability of the electricity supply pattern over time,[1] and the ability of renewables to provide capacity on electricity networks. These characteristics are resource specific, and it is important to understand the properties of different renewable resources. In the same way as there are large differences in the operation of different conventional resources (e.g. coal, gas and nuclear), there are differences in the characteristics of renewable sources of electricity generation.

This chapter commences with an analysis of the current make-up of renewable electricity generation in the UK and the expected direction of the sector's evolution in terms of energy contributions from different resources. Key characteristics of three renewable energy resources in the UK – wind power, wave power and tidal stream power – are presented, and the chapter concludes with an analysis of recent claims regarding the capacity credit of renewables (in particular, wind power).

RENEWABLE ELECTRICITY GENERATION IN THE UK

Renewable electricity production in the UK utilizes a wide range of energy resources as fuels for the generation of electricity, from methane (from landfill gas), biomass and waste, to wind and (in the future) waves and tidal currents.

Accompanying this range of renewable resources is a range of resource-specific characteristics: for example, the ability to store biomass and waste before conversion to electricity provides a high degree of control over the timing of electricity production. By contrast, energy from tidal currents is very predictable; however, it is not possible to control the timing of tides and, therefore, the pattern of electricity production at individual tidal sites.

By assessing the underlying resource properties, renewable electricity generators can be broadly classified into two categories: those such as landfill gas, biomass or hydro, where their pattern of electricity generation can be controlled (dispatchable capacity); and those such as wind, wave, tidal and solar resources, where the pattern of generation is dependent upon external factors (non-dispatchable or variable-output capacity). The ability to control the electricity generation of dispatchable renewables is analogous to the operation of conventional generating capacity, such as coal, gas and nuclear capacity. Variable generation represents a departure from this conventional operation of generating capacity since it is the speed of the wind (or height and period of the waves, velocity of the tidal current, etc.) that determines the level of generation at any particular time.

The UK's renewable electricity portfolio currently includes a combination of both dispatchable and variable-output renewables, with dispatchable renewables being responsible for over 75 per cent of renewable electricity generation (see Figure 3.1). While wind power, both onshore and offshore, provides almost all variable generation from renewables, it is currently responsible for less than one quarter of total renewable energy production. Given that renewable electricity (including large-scale hydro) accounted for around 4 per cent of UK electricity, variable-output renewables were responsible for less than 1 per cent of UK electricity generation in 2005 (DTI, 2006a).

The contribution of variable-output renewables to renewable electricity generation, and to overall electricity supply in the UK, is expected to increase into the future. The Renewables Obligation currently supports a renewable electricity target of 15.4 per cent electricity generation by 2015, with support for renewable electricity generation continuing until 2027 (UK Parliament, 2006). However, it has been proposed that the Renewables Obligation be expanded to 20 per cent of electricity generation by 2020 and potentially modified to include targeted support for under-represented resources through a 'banding' mechanism (DTI, 2006b). Forecasts for 2020 suggest that wind power will represent around 70 per cent of renewable electricity generation (see Figure 3.2), while the inclusion of electricity generation from wave and tidal stream resources will see total variable-output renewable electricity supply account for over three-quarters of renewable electricity, and around 15 per cent of total electricity generation (The Carbon Trust and DTI, 2003).

Given such scenarios for future renewable electricity generation, and the desire of government to target funding at emerging renewable technologies, it will become increasingly important for the electricity supply properties of variable-output renewables to be known. And while wind power is forecast to dominate this sector of renewable generation, the emergence of other variable-

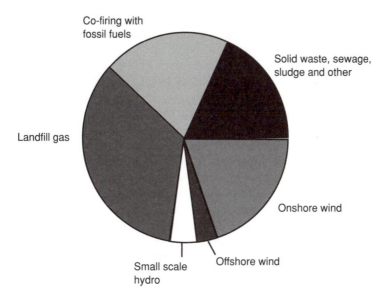

Source: DTI (2006a)

Figure 3.1 *Contribution of different renewable resources to total renewable electricity generation in the UK in 2005 (excludes large-scale hydro)*

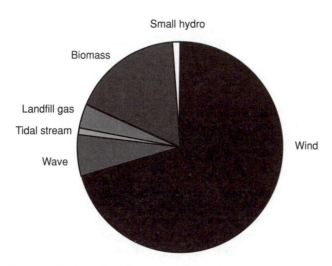

Source: The Carbon Trust and DTI (2003)

Figure 3.2 *Forecast contribution of different renewable resources to total renewable electricity generation in the UK in 2020 (excludes large-scale hydro)*

output technologies suggests that the characteristics and interaction of a diversified portfolio of variable-output renewables will be valuable in understanding their interaction with demand patterns and network integration.

CHARACTERISTICS

The electricity production characteristics of variable renewables reflect the availability, or change in availability, of the underlying 'fuel' being used to generate electricity. In the UK, wind power is expected to develop rapidly into the dominant method for supplying renewable energy, meaning that the UK wind resource will be important in determining the characteristics of electricity supplied by this resource. Much of the network modelling work carried out on the UK electricity network has reflected this expectation by evaluating the impact of a 20 per cent wind power scenario (Dale et al, 2004; Mott MacDonald, 2004; Anderson, 2006). With the emergence of wave and tidal power systems, an understanding of the underlying characteristics of these resources will also be required.

This chapter presents an overview of the patterns of energy production from UK wind, wave and tidal stream renewable resources, the variability of supply associated with each resource, the relationship between electricity generation and demand patterns, and the impact that large-scale electricity generation could have on the net electricity demand pattern to be met by conventional generators.

Patterns and variability of renewable energy availability

The availability of renewable energy varies over time, with location and with the resource being exploited. The pattern of energy availability will also change depending upon the time scale being considered. For example, tidal stream energy shows very little change in energy production from year to year or month to month; however, large changes in output can occur from hour to hour. Contrast this with wave power, where output tends to change less than tidal stream at the hourly level, but there are significant differences at the monthly time scale – although these are small at an inter-annual time scale (see Table 3.1).

The variability of a resource is also affected by the level of diversification in its development. By exploiting a resource such as wind power at a range of sites, the variability exhibited by the aggregate output from all wind developments would be reduced. This arises as the correlation between power output patterns at different sites decreases as the distance between the sites increases (Giebel, 2000; Sinden, 2007). With different sites experiencing different wind conditions at any one time, changes in power output at one site (such as an increase in power output in response to increasing wind speed) may be partially offset by a decrease in power output at another site on the same network.

Table 3.1 *Relative variability of wind, wave and tidal resources over different time periods*

Resource	Annual	Monthly	Hourly
Wind	Low	Medium (over the year)	Low
Wave	Low	Medium/high (over the year)	Very low
Tidal stream*	Very low	Very low	Very high

Note: * The spring-neap tide cycle results in large changes in power output over a two-week period. However, these changes are smoothed at the monthly time step, resulting in low variability in tidal stream electricity production from month to month.

This smoothing characteristic is particularly apparent at short time scales (such as hour-to-hour generation patterns): diversification is unlikely to affect the seasonal variability of wind power unless the area over which developments can occur is exceptionally large. At the monthly time scale, the long-term availability of renewable resources differs significantly between wind, wave and tidal stream resources (see Figure 3.3). Wave power shows the greatest extremes in monthly output, with energy delivery peaking in January and February, and reaching a minimum in July. Monthly energy production from wave power over winter is typically more than five times the average monthly energy generation during summer months. Wind power follows a similar seasonal pattern – an expected result, given that ocean waves form as a result

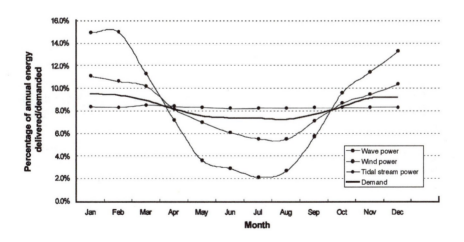

Figure 3.3 *Percentage of annual electricity generation by wind, wave and tidal power systems occurring during each month, with long-term UK electricity demand distribution shown for comparison (normalized to account for differing days per month)*

of the action of wind. However, the seasonal differences in wind power are not as extreme as those found in the wave record. Energy production from wind power is again dominated by the winter months, with minimum production occurring during the summer months – typical winter month energy production is around 1.8 times higher than summer month production. Tidal stream energy shows a very different distribution of energy production over the year, with average monthly energy production essentially constant across the year.

The monthly trend in electrical energy demand for the UK (see Figure 3.3) shows some similarity to the distribution of energy production from wind and wave power systems across the year, with higher demand in winter months and lower demand during summer months. While this trend is encouraging, electricity networks need to keep electricity supply and demand in constant balance. The section on 'Renewable electricity supply and demand patterns' considers the relationship between electricity demand and energy production from variable-output renewables.

Hour-to-hour variation in output

A key aspect of variable-output renewable resources is the rate of change in power output over time. The rate of change in output of a wave, wind or tidal stream electricity generator depends upon two factors – the sensitivity of the device to changes in environmental conditions and the degree to which environmental conditions change.

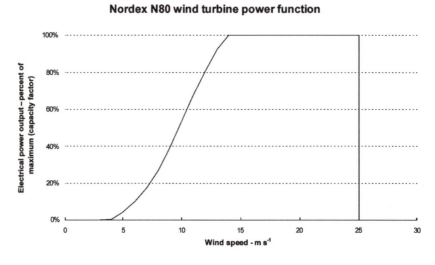

Source: Nordex (2006)

Figure 3.4 *Power transform curve for a typical large wind turbine*

Consider the power transform curve for a Nordex N80 wind turbine (see Figure 3.4): power output is most sensitive to changes in wind speed between 8 metres per second (m s⁻¹) and 12m s⁻¹, while outside this range the sensitivity of the wind turbine to changes in wind speed reduces. Changes in wind speed below 4m s⁻¹ do not result in any change in power output since there is no energy production; at the other extreme, changes in wind speed between 14m s⁻¹ and 25m s⁻¹ do not result in changes to output as the device operates at rated capacity across this wind speed range. Almost 90 per cent of hours experience wind speeds between 4ms⁻¹ and 14ms⁻¹, the speed range associated with changes in wind power output. While the development of a diversified wind power sector will slightly reduce the sensitivity of power output to changes in wind speed (Sinden, 2007), it remains the case that changes in wind speed within this range will result in changes in power output. The second component of power-output variability is the degree to which the resource availability changes – for example, while a change in wind speed from 8ms⁻¹ to 12ms⁻¹ would result in a significant increase in power output from a wind turbine, it is the frequency with which such changes occur that ultimately determines the variability of the electricity supply.

The power output from wave and tidal stream energy devices is similarly related to the sensitivity of the device and the degree to which the underlying resource (i.e. wave height and period for wave power, or current velocity for tidal stream power) fluctuates from hour to hour; however, the hour-to-hour variability of tidal stream power output is highly dependent upon the location and degree of resource development. In some regions, such as the Channel Isles, the characteristics of the tidal stream resource allow for significant smoothing of the power production pattern since the timing of peaks and minimums in tidal stream current velocities varies between the different sites (see Figure 3.5). However, other regions, such as the Pentland Firth, offer very limited opportunities for diversification to smooth output variability as the power production patterns from the different sites within the region are all highly correlated (Sinden, 2005).

The difference in power output from one hour to the next for wind and wave power output at the one-hour time step is shown in Figure 3.6 (note that tidal stream power is not included in this graph due to the site-specific nature of variability for this resource). Wave power shows a lower level of variability than wind power in the UK, with a higher frequency of little or no change in power output occurring from one hour to the next. The graph in Figure 3.6 has been truncated to show the frequency of hourly power output changes between ±20 per cent of installed capacity – for wind power, 99.95 per cent of all hourly changes in power output were less than 15 per cent of the installed wind power capacity, with 46 per cent of hourly changes being equivalent to less than 1 per cent of the installed wind power capacity. Wave power showed a greater concentration of low-change hours, with 61 per cent of changes in power output being equal to less than 1 per cent of the installed wave power capacity, and with 99.975 per cent of all hourly changes in power output being between ±20 per cent of installed capacity.

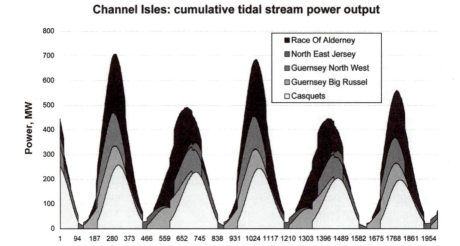

Note: Results are based on full development of the identified tidal resource at each site (see Black and Veatch, 2005).

Figure 3.5 *Relative timing and contribution to renewable output from individual tidal stream sites in the Channel Isles*

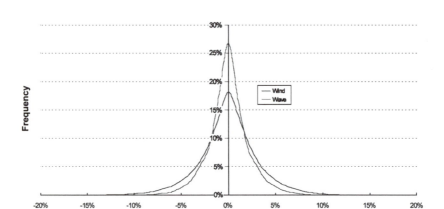

Note: The graph is truncated to ±20 per cent of installed capacity; this shows more than 99.95 per cent of all hourly changes.

Figure 3.6 *Variability of wind and wave power output at the one-hour time step*

RENEWABLE ELECTRICITY SUPPLY AND DEMAND PATTERNS

Relationship between renewable energy availability and demand

While there is a natural correlation in the UK between the availability of wind and wave power and electricity demand at a monthly time scale (see Figure 3.1), this degree of aggregation removes the inter-day variability that is present in both demand and supply. By determining the availability of renewable resources at an hourly time scale and comparing this availability to hourly electricity demand, a more detailed view of the availability of renewables during different demand periods can be gained.

In the UK, wind and wave power deliver more energy during times of high electricity demand, with periods of low electricity demand typically being associated with significantly lower wind power output (see Figure 3.7). Wave power shows higher average energy production during peak demand periods than does wind power; however, this energy production tends to drop quickly in hours with lower demand. Average energy production from tidal stream power (see Figure 3.7) is essentially uniform across all demand hours. This tendency for different average energy production levels at different demand levels is highlighted in Figure 3.8, which shows the degree to which renewable production from wind and wave power, together with demand, deviate from the long-term average.

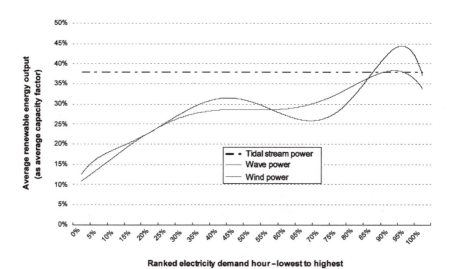

Ranked electricity demand hour – lowest to highest

Note: Average annual capacity factor is ~28 per cent for both wind and wave power, and 38 per cent for tidal stream power.

Figure 3.7 *Average wind and wave power availability (as capacity factor) at different levels of demand*

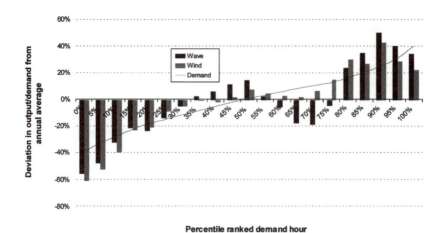

Figure 3.8 *Deviation of wind power, wave power and demand from annual average production (and demand) levels*

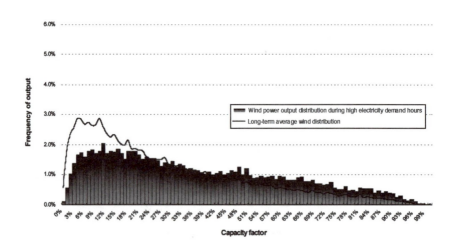

Source: Sinden (2007)

Figure 3.9 *Distribution of hourly wind power output during peak electricity demand*

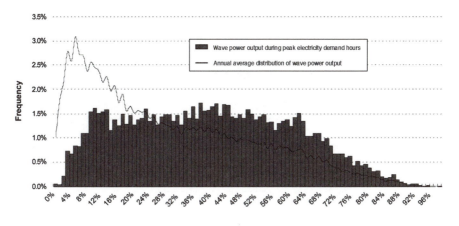

Wave power output (as capacity factor)

Source: Sinden (2007)

Figure 3.10 *Distribution of hourly wave power output during
peak electricity demand*

These results do not mean that all high demand hours will experience higher
than average power output from wind and wave resources as there remains
considerable variability in the hourly output of wind power during high electricity demand hours. The actual power output that may be experienced from
these resources varies across a wide range. However, during high electricity
demand hours (the top 20 per cent of demand hours), the occurrence of low
power-output events is less than the long-term average, while medium and high
power events occur more frequently than the long-term average (see Figures
3.9 and 3.10).

Impact of renewables on demand variability

While the long-term availability of wind and wave power is broadly correlated to changes in electricity demand, output variability from these resources,
together with tidal stream power, will affect the net electricity demand pattern
to be met by conventional generators. At the hourly level of demand fluctuation, the development of significant renewable electricity generating capacity
will increase the hour-to-hour change in demand to be met by conventional
generators (see Figure 3.11). This impact varies both with the size of the
renewable resource development and with the composition of the renewable
capacity: where renewables deliver 5 per cent or less of total energy to the
electricity network over a year, there is little impact on net demand variability,

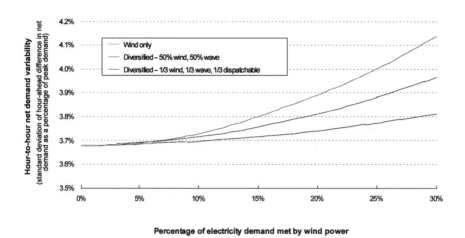

Figure 3.11 *Hour-to-hour variability in demand to be met by conventional capacity with an increasing proportion of renewable electricity generation*

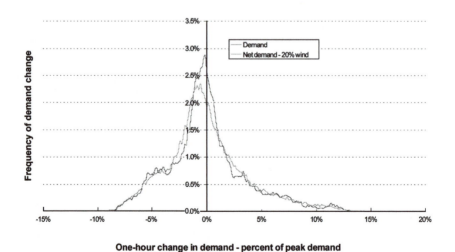

Figure 3.12 *Impact of 20 per cent of electricity demand met by wind power on the magnitude and frequency of demand changes to be met by conventional capacity*

irrespective of the composition of renewable resources. However, increasing the total renewable electricity contribution beyond 5 per cent results in both increased net demand variability and a clear distinction between the level of impact arising from different portfolios of renewables. A wind-only scenario results in the highest net-demand variability, while a portfolio comprising equal energy production from wind, wave and other dispatchable renewable generation has the lowest impact on net demand variability. Under a 20 per cent renewable energy scenario, a diversified wind–wave dispatchable renewables portfolio would have less than one third the impact on the hour-to-hour change in demand compared with that seen for a wind-only scenario, while a combined wind–wave portfolio would result in a 37 per cent reduction in net demand variability compared to the wind-only scenario (see Figure 3.11).

These changes in demand variability would be most commonly felt as a reduction in the proportion of hours each year experiencing very low change in demand levels from one hour to the next (see Figure 3.12). In the example shown, the reduced frequency of small changes (±2 per cent of peak demand) from one hour to the next is balanced by the increased frequency of hourly changes in the range of ±2 to 4 per cent of peak demand.

The results presented in this chapter quantify some of the impacts that the development of variable output renewables would have on electricity networks. They also demonstrate the importance of resource type in assessing the impacts of renewables – while wave and wind power both show a similar distribution of energy production with demand level, a combined wind–wave renewables portfolio will have a lower impact on existing electricity-demand patterns than a wind-only scenario.

THE ROLE OF WIND POWER IN PROVIDING CAPACITY ON ELECTRICITY NETWORKS

The total available capacity on a network should always exceed the peak demand met by the network, and this difference is known as the plant margin (or backup). A plant margin is required to ensure that demand will be met to a specified level of reliability: as no generating capacity is 100 per cent reliable, the provision of more capacity than is needed to meet demand is essential for the long-term reliable performance of the electricity network. The size of the plant margin is related to the electricity-demand pattern, the reliability of the generating plant, the required overall reliability level of the network, and other network parameters.

While dispatchable renewables already provide capacity on electricity networks, the impact of the variable electricity supply profile of renewables such as wind and wave power needs to be considered. There has been extensive discussion regarding the plant margin that is required on the UK electricity network, and the degree to which renewable generating capacity provides capacity credit (the amount of conventional generating capacity that can be removed from a network due to the addition of renewable generating capacity, while maintaining the same level of security of supply) on a network. In assessing the capacity credit of renewables, two key principles apply:

1   *Maximum conventional capacity requirement.* With the addition of renewable generating capacity, the total conventional capacity requirement can only stay the same or fall; it cannot increase. For example, if wind power is added to a network that currently requires 84 gigawatts (GW) of plant to meet demand; the total conventional plant requirement will not exceed 84GW because of the development of wind power.
2   *Actual conventional capacity requirement.* The actual conventional plant requirement will fall if the performance of the renewable capacity contributes to operational security. For example, the addition of 1GW of dispatchable renewable capacity such as landfill gas would result in 1GW less conventional plant being required since it has the same (or better) reliability of supply than the conventional plant that it is replacing.

In the example above, the landfill gas capacity would have a capacity credit of 100 per cent since it has substituted directly for an equivalent amount of conventional capacity on the network. Variable-output renewables such as wind power have a limited capacity credit as their probability of generation at times of peak demand is lower than that for conventional or dispatchable renewable capacity. Two recent studies have attempted to clarify the question of whether wind power provides capacity on electricity networks: one reviewing 29 separate studies (UKERC, 2006, summarized in Chapter 4 of this volume) and the other reviewing over 50 studies (Giebel 2006). Both reviews come to the same conclusion: every study into the capacity credit of wind power shows that it does have a capacity credit. There are, of course, differences in the absolute amount of capacity credit attributable to wind power; however, this is not surprising given that the capacity credit of wind power (and other renewables) is affected by the assumed size of the development in relation to the network, the demand pattern to be met and the site-specific characteristics of the resource.

Determination of capacity credit

The capacity credit of renewable capacity on an electricity network is an outcome of reliability assessments typically carried out through a technique known as loss of load probability (LOLP) analysis. Under this method, the probability that a given demand will be met by the available generating capacity is determined. In recent studies (Dale et al, 2004; Anderson, 2006), it has been assumed that conventional generators have an availability of around 85 per cent. By statistically combining these individual generating units, a distribution of available capacity is determined and the probability of demand exceeding available capacity (i.e. loss of load) is calculated.

An example of two future electricity networks, one comprised solely of conventional generators and one where 20 per cent of the energy demand of the network is met by wind power, is presented by Dale et al (2004). Under the first scenario, a future UK electricity network has a peak demand of 70GW and energy demand of 400 terawatt hours per year (TWh y^{-1}); demand is met by generators with an availability of 85 per cent; and in order to achieve a specified reliability of supply, around 84GW of plant are needed in total (as determined by a probability analysis). A second alternative future UK electricity network has 20 per cent (80TWh y^{-1}) of its electricity coming from wind power, generated by 26GW of wind capacity (with a 35 per cent annual capacity factor), with 79GW of conventional capacity required on the network. Thus, the development of wind power on the network has led to a 5GW reduction in conventional capacity requirement – a wind capacity credit of 5GW – while achieving the same level of security of supply.

Wind power and backup

Concern has been expressed regarding 'backup' or the additional plant that is needed to offset the variability of renewables, such as wind power. Some authors have suggested that backup equal to 65 per cent of the installed capacity of wind power (PB Power and RAE, 2003), or even 100 per cent of the capacity of wind power (Laughton, 2002; Fells Associates, 2004), would be required to be installed in response to the development of wind power on an electricity network.

One method of quantifying the 'backup' requirement of wind power is to compare the capacity credit of wind power to the potential maximum reduction in conventional capacity due to the development of the wind resource. Following on from the scenario presented by Dale et al (2004), a 20 per cent wind power scenario would yield around 80TWh y^{-1} of wind energy: on an equal-generation basis, 80TWh y^{-1} would be generated by 10.7GW of conventional plant. Ideally, the wind plant would substitute directly for the conventional plant, meaning that around 73.3GW of conventional plant would be needed. However, this does not happen. Instead, around 79GW of conventional plant are required to achieve the same level of reliability (security of supply). This difference of 5.7GW represents the additional capacity required due to variability of wind power.

In this example, the wind capacity has an additional backup requirement due to the variability of wind power of around 22 per cent (5.7GW expressed as a percentage of the 26GW of installed wind capacity). By contrast, a 65 per cent backup scenario suggests 16.9GW of additional generating capacity would be required to accompany a 26GW wind power development, while a 100 per cent backup scenario would see an additional 26GW of thermal plant constructed due to the variability of wind power. Following the above example, a 100 per cent backup scenario would see a conventional system that requires 84GW of generating capacity to reliably meet demand have its conventional capacity requirement reduced by 10.7GW (thermal capacity equivalent of the new wind capacity in energy production), and in its place would be developed 26GW of wind power and an additional 26GW of conventional capacity as backup for the wind capacity. The result is a network with 99.3GW of conventional capacity (84GW – 10.7GW + 26GW) plus

26GW wind capacity, a curious result given that the maximum conventional capacity requirement would not exceed the original 84GW even if the 26GW of wind power produced no energy for the entire year. These different backup scenarios are compared against the conventional capacity-only scenario in Figure 3.13.

Other variable output renewables

The discussion of capacity credit and backup presented here uses wind power as the sole source of renewable electricity generation on the network. However, this is not the case, at present, for the UK, and while wind power is forecast to dominate renewable electricity production in the medium term (The Carbon Trust and DTI, 2003), non-wind renewable capacity will continue to contribute to aggregate UK renewable electricity generation. One recent study (ILEX and Strbac, 2002) has presented an analysis of combined wind and biomass renewable capacity on the UK electricity network, while results by Sinden (2006) suggest that it is possible to improve the capacity credit of variable output renewables by developing a diversified portfolio of renewable resources. Given the diversity of renewable resources available for development, and current indications that UK renewable-energy support policies may directly target additional types of renewable resources, the role of diversity in the renewables sector is likely to become an important area of debate.

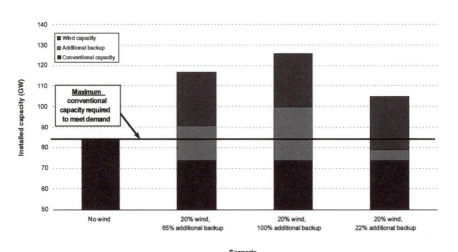

Note: The 65 and 100 per cent backup scenarios result in a higher conventional capacity requirement than the original network without wind power.

Figure 3.13 *Total capacity implication for a conventional capacity network and three alternative estimates of the conventional capacity required due to wind power*

CONCLUSIONS

The characteristics of renewable resources are central to the large-scale integration of renewables within electricity networks. These characteristics are both site specific and resource specific, with markedly different power production patterns arising from different resources and locations. This chapter has presented a summary of characteristics relating to the UK wind, wave and tidal stream resource, including the relationship between long-term resource availability and electricity demand patterns, and the effect that large-scale development of renewables may have on the residual demand pattern to be met by conventional capacity.

The role that renewables can play in reducing the conventional capacity requirement of an electricity network has been discussed with reference to studies that demonstrate that wind power has the effect of reducing the conventional capacity requirement of the network. Alternative views of the impact of large-scale renewable development on capacity requirements have been presented; however, it was argued that these do not provide a reliable assessment of the impact of wind power on network capacity requirements. Finally, it was argued that future analyses of renewable resources on electricity networks may have to increasingly take account of the different characteristics of the renewable resources themselves.

ACKNOWLEDGEMENTS

Analyses presented in this chapter are based on extensive historic and model data, and the author acknowledges the generosity of the UK Met Office, British Atmospheric Data Centre, Prodman Oceanographic Laboratory and National Grid for making data and models available for this research.

NOTE

1   Variability has sometimes been referred to as 'intermittency'; however, this terminology may mislead as it implies that a resource is either present or absent. For resources such as wind power, it is changes in wind speed rather than the presence or absence of wind that are the key feature of the resource. The term intermittency may be more applicable to conventional generators, particularly base-load generators such as nuclear, which typically exhibit a presence (operation at rated capacity) and absence (zero output due to maintenance or mechanical failure) pattern of electricity generation.

REFERENCES

Anderson, D. (2006) *Power System Reserves and Costs with Intermittent Generation*, UKERC, London, p47
Black and Veatch Corporation (2005) *Phase II UK Tidal Stream Energy Resource Assessment*, The Carbon Trust, London

Dale, L., Milborrow, D., Slark, R. and Strbac, G. (2004) 'Total cost estimates for large-scale wind scenarios in the UK', *Energy Policy*, vol 32, p8

DTI (UK Department of Trade and Industry) (2006a) *Digest of United Kingdom Energy Statistics. National Statistics*, The Stationery Office, London, p226

DTI (2006b) *The Energy Challenge*, DTI, HM Government, London, p218

Fells Associates (2004) *Submission to the House of Lords Science and Technology Committee Inquiry into Renewable Energy*, House of Lords, HL Paper 126-II, London

Giebel, G. (2000) 'On the benefits of distributed generation of wind energy in Europe', *Vom Fachbereich Physik*, Carl von Ossietzky Universität Oldenburg, Oldenburg, p117

Giebel, G. (2006) *Wind Power Has a Capacity Credit*, Risø National Laboratory, Technical University of Denmark, Roskilde

ILEX and Strbac, G. (2002) *Quantifying the System Costs of Additional Renewables in 2020*, ILEX Consulting, Oxford and Manchester Centre for Electrical Energy, University of Manchester Institute of Science and Technology, p136

Laughton, M. (2002) 'Renewables and UK grid infrastructure', *Power in Europe*, vol 383, no 1, pp9–11

Mott MacDonald (2004) *Renewable Network Impact Study. Annex 4: Intermittency Literature Survey and Roadmap*, The Carbon Trust and the Department of Trade and Industry, London

Nordex (2006) *Nordex N80/2500 Product Data Sheet*, accessed 28 October 2006, www.nordex-online.com/fileadmin/MEDIA/Produktinfos/EN/Nordex_N90_N80_Produktbroschuere_EN.pdf

PB Power and RAE (2003) *The Cost of Generating Electricity*, Royal Academy of Engineering, London, p60

Sinden, G. (2005) *Variability of UK Marine Resources*, Environmental Change Institute and The Carbon Trust, London, p93

Sinden, G. (2006) *Diversified Renewable Energy Resources*, Environmental Change Institute and The Carbon Trust, London, p42

Sinden, G. (2007) 'Characteristics of the UK wind resource: Long-term patterns and relationship to electricity demand', *Energy Policy*, vol 35, no 1, pp112–127

The Carbon Trust and DTI (2003) *Renewables Network Impact Study. Annex 1: Capacity Mapping and Market Scenarios for 2010 and 2020*, Renewables Network Impact Study, The Carbon Trust, London, p61

UK Parliament (2006) *The Renewables Obligation Order: Schedule 1, Statutory Instrument 2006 No 1004*, The Stationery Office, London

UKERC (UK Energy Research Centre) (2006) *The Costs and Impacts of Intermittency: An Assessment of the Evidence on the Costs and Impacts of Intermittent Generation of the British Electricity Network*, UK Energy Research Centre, Imperial College, London, p112

4

The UK Energy Research Centre Review of the Costs and Impacts of Intermittency

Robert Gross, Philip Heptonstall, Matthew Leach,
Jim Skea, Dennis Anderson and Tim Green

Introduction

This chapter is based upon work undertaken by the UK Energy Research Centre (UKERC) and reported in *The Costs and Impacts of Intermittency: An Assessment of the Evidence on the Costs and Impacts of Intermittent Generation on the British Electricity Network* (Gross et al, 2006). The study sought to provide a detailed review of the current state of understanding of the engineering and economic aspects of intermittent/variable generation.[1] The approach adopted was inspired by a range of techniques referred to as *evidence-based policy and practice*, including the practice of *systematic review,* common in policy areas such as healthcare and education. These seek to inform policy by examining the research data in a transparent and replicable way. There are limitations on the application of such techniques to the energy policy arena (Sorrell, 2007), but a systematized and transparent approach to evidence-gathering offers considerable benefits, particularly where the topic is controversial.

A systematic search for every report and paper related to the costs and impacts of intermittent generation was undertaken. This revealed over 200 reports and papers on the subject, each of which was categorized and assessed, to draw out the methodologies and data employed, and the relevance to the UK situation. UKERC also sought consensus through the convening of an expert group with diverse views and consultation with stakeholders.

A degree of confusion appears to surround the use of terminology, and a number of misconceptions are regularly aired in the mainstream media. The UKERC report therefore also explains the principles that underpin the integration of intermittent renewables and aims to explore and explain popular misconceptions.

This chapter first provides an overview of the key concepts of relevance to power system reliability and operation, and explains what changes when intermittent generation is introduced. Some popular misconceptions are revisited. The final section presents and discusses the quantitative findings of the UKERC review.

POWER SYSTEM RELIABILITY AND OPERATION

Managing electricity networks without intermittent generation

The impacts of intermittency cannot be considered in isolation from the main principles of electricity system operation. Electricity demand changes continuously. It fluctuates from second to second and goes through very large swings over a few hours. While the general direction of demand each day and at different times of the year is well understood, demand can still change unexpectedly. Large power stations also experience occasional unexpected faults.

Currently, in the UK, matching of demand and supply is handled through the operation of regulated markets. The UK market (excluding Northern Ireland) is governed by the British Electricity Trading and Transmission Arrangements (BETTA). The system operator (National Grid) works with the market mechanisms and, through a range of specific interventions, stimulates increased electricity generation as demand increases, as well as reductions as demand falls.[2] A thoroughgoing explanation of how the UK system is balanced and how reliability is maintained is provided in Gross et al (2006). We highlight some key facets relevant to the integration of renewable resources below.

Short-term system balancing requirements

A range of services referred to as system balancing reserves are purchased by the system operator in order to help deal with unexpected short-term fluctuations (seconds to hours). These can be caused by either demand changes or faults at power stations or power lines. The amount of reserve needed to handle unpredicted short-term variations is calculated or simulated using statistical principles. The objective is to ensure that reserves are available that can deal with almost all the unpredicted fluctuations envisaged.

Longer-term system reliability requirements

In addition to the short-term reserves, a larger 'system margin'[3] of maximum possible supply over peak demand is provided for when planning the development of a power system. UK practice before liberalization was to plan for and invest in installed capacity approximately 20 per cent larger than expected peak demand. Current practice by the system operator is to monitor and report on this margin, and to allow the market to respond to any perceived lack of system margin.

The aim is to ensure that a specific measure of reliability is sustained, with only a small risk of demand being unmet. The measures most often used when assessing the impact of generation reliability on customers are related to loss of load. Loss of load probability (LOLP) expresses the risk that some load will need to be shed because insufficient generation is present. This is expressed as a percentage corresponding to the maximum number of years per century in which load shedding may occur (e.g. the LOLP of the pre-privatized electricity system in Great Britain was planned not to exceed 9 per cent: nine loss-of-load events per century).

The distinction between the system margin required for longer-term reliability and reserves required for short-term balancing is illuminated by the comparative size of the two quantities. In the UK, balancing reserves are purchased by the National Grid Company (the transmission system operator) and comprise around 2.5 gigawatts (GW).[4] System margin is much larger than dedicated reserve and is not contracted for: National Grid's indicative level of adequate system margin is around 20 per cent above peak demand, or 12GW to 14GW.[5]

What changes when intermittent generation is added?

While all plant will suffer occasional outages, intermittent renewables fluctuate to a much greater degree, and this may increase the need for balancing by the system operator. Their output also may or may not be available during peak demand periods; the extent to which it can be relied upon at peak varies according to resource type and location. Availability during demand peaks affects the amount of system margin needed to maintain a given measure of reliability. Hence, both system balancing and system reliability can be affected by the introduction of renewable sources.

Impact on short-term system balancing requirements

Additional short-run fluctuations in output will increase the use of reserve capacity – for example, the utilization of automatic controls on the output of conventional power stations. It may also be necessary to have more part-loaded plant running that can rapidly increase or decrease its output. Three interrelated factors determine the amount of extra reserve required when intermittent generation is added to the system:

1   The extent of unpredicted variations in intermittent plant output, which has two aspects: first, how *rapidly* the outputs of different types of intermittent plant will fluctuate; and, second, the possible scale of *total* system-wide changes in a given period. This requires a representation of the aggregated behaviour of individual intermittent plants, based on weather data, size of units, inertia, and the scope for 'smoothing' of outputs – for example, by geographical dispersion. This data provides an indication of the variability of intermittent supplies.
2   How accurately fluctuations over the seconds-to-hours time scale can be forecast. The more accurate the forecasting, the greater the opportunity to use (lower-cost) *planned* changes as opposed to holding reserve plant in readiness, which can be costly and less efficient. In market terms, the effects of predicted fluctuations can be contractually committed prior to 'gate closure' (which in the UK is the point one hour ahead of real time when bilateral electricity contract positions must be notified to the system

operator). This should permit the market to reveal the most cost-effective means of managing these variations. Again, it is the prediction accuracy of total aggregated intermittent generation that is relevant; forecasting for a large amount of distributed resources reduces forecast errors.

3   How the timing of demand variations compares with that of intermittent output. If, for example, load and output are highly correlated, their variations may act to cancel each other out and to reduce reserve requirements.

In all cases, analysis requires a statistical treatment of both demand and intermittent generation since we are dealing with *probabilities* rather than determinate functions.

Two factors are notable: first, that even for a relatively unpredictable intermittent source such as wind power the standard deviation[6] (and the variance)[7] of fluctuations in the period from minutes to a few hours is relatively modest. This is because there is considerable smoothing of outputs in the sub-hourly timeframe and considerable prediction accuracy over a few hours. Second, variance of intermittent fluctuations must be combined statistically with the variance of demand and conventional supply.

Impact on longer-term system reliability requirements

The extent to which intermittent generation can replace thermal plant without compromising system reliability is referred to as its capacity credit, where a credit of 100 per cent denotes one-for-one substitution without loss of reliability, and 0 per cent indicates that the intermittent source can displace no conventional capacity. It is important to note that capacity credit is a derived term because it can only be calculated in the context of a more general assessment of reliability across peaks.[8]

It might be thought that intermittent plant cannot contribute to reliability at all since, in most cases, we cannot be certain that it will be available at any specific time, including at system peaks. However, there is a possibility that any plant on the system will fail unexpectedly and reliability is always calculated using probabilities. Intermittent plant can contribute to reliability provided that there is some probability that it will be operational during peak periods. The key determinants of capacity credit are as follows:

- *The degree of correlation between demand peaks and intermittent output:* the greater the correlation, the greater the capacity credit. For example, photovoltaics (PV) have zero capacity credit in the UK because demand peaks occur during winter evenings, when it is dark. But PV can have a high capacity credit in sunnier regions where demand peaks are driven by air-conditioning loads that are highly correlated with PV output.
- *The average level of output:* a higher level of average output over peak periods will tend to increase capacity credit. Taking UK wind as an example, there is little correlation between wind output and demand in the hourly and daily timeframes. However, wind farm outputs are generally higher in winter than they are in summer. For this reason, analysts use winter-quarter wind output to calculate capacity credit.

- *The range of intermittent outputs:* where demand and intermittent output are largely uncorrelated, a decrease in the range of intermittent output levels will tend to increase capacity credit because the variance decreases. More consistent wind regimes decrease variance and increase capacity credit. Geographical dispersion of plants can smooth outputs and decrease overall variation, as can increasing the variety of types of intermittent plant on a system.

Capacity credit is determined by considering the total variance of both supply and demand, including intermittent options on the supply side, and then comparing this to a 'without intermittency' case. Because the variance at peak demand is larger than for conventional stations, the capacity credit of intermittent sources tends to be lower than their availability, and at larger penetrations is also less than capacity factor. The range of capacity credits, the reasons for the range, and the implications for system costs are explored in the section on 'Quantitative findings on impacts and costs'.

Misconceptions and sources of controversy

Terminology

Terminology can give rise to two important areas of misunderstanding. The first problem is often associated with the use of terms such as 'backup' or 'standby' generation. Confusion arises when these terms become linked with the idea that intermittent sources need dedicated backup. This is incorrect for the following reasons:

- Actions to manage short-term fluctuations and to maintain reliability of electrical networks should be assessed on the basis of plants interconnected and operated as a *system*. Dedicated backup is not required (in the same way that individual conventional power stations do not require dedicated backup to cope with unexpected failures). Rather, intermittent plants may increase the amount of reserves and response needed for balancing the system, may affect the efficiency of other plants, and may increase the amount of capacity on the system that is required to maintain reliability.
- The additional actions needed to manage fluctuations from intermittent plants are also affected by the nature of fluctuations resulting from demand changes and the reliability of conventional stations on the system (the extent being dependent upon their relative magnitudes and correlations). Failure to assess these requirements in a systemic fashion would only be consistent if it were applied to all generation since all experience unplanned outages.

The second problem is associated with the use of the term 'reserves'. The term is used for quite different types of function on different time scales. Two broad categories of usage can be found in the literature:

1   *Reserves* has a strict and narrow sense, restricted to the requirements for fast responding reserves for short-term *system balancing* that are contracted for by the system operator. These are the only reserve services that the system operator directly purchases in the UK.

2   A broader definition also encompasses the additional capacity that may be required to ensure *reliability* when viewed from a long-term, or planning, time horizon. *System margin* is the current terminology used to refer to this capacity, and in Britain there is no mechanism for direct procurement of system margin. Yet, historically, under centrally planned systems, capacity over and above peak demand has also been referred to as capacity reserves.

This can mean that comparisons are drawn between studies that are using the term differently. For example, some studies of the cost of intermittency in fact only quantify the cost of additional system balancing – the capacity to maintain reliability may be neglected or may not be directly addressed. This may give rise to a 'reserve cost' estimate that understates the full cost of intermittency. However, where the term 'reserves' is used to refer to *both* capacity provision to maintain reliability *and* short-term reserves, this, too, can create confusion since it leads to cost estimates considerably larger than those directly attributable only to the reserve services actually purchased by the system operator.

Misconceptions

Two related assertions that receive regular airings in the mainstream media are paraphrased below:[9]

1   'Wind turbines only operate 30 per cent of the time; therefore, we must provide 70 per cent backup.'
2   'Wind turbines need backup, so they don't save any carbon dioxide (CO_2).'

Both these assertions are incorrect. In both, the use of the term 'backup' may, in itself, give rise to misunderstanding for the reasons outlined above. But irrespective of terminological issues, the assertions are in error for the following reasons:

- The former assertion confuses the *capacity factor* that might be achieved by a typical UK wind farm (which would, indeed, be around 30 per cent on an annual basis in a UK location with good wind conditions) with the amount of *time* during which it is operational. In fact, most wind turbines will be operational for around 80 per cent of the time – but usually operating at less than their rated capacity. This is because the rated capacity of a wind turbine is its *maximum* output, which is typically associated with wind speeds from in excess of 11 metres per second (m s^{-1}) to 15 m s^{-1} (40km h^{-1} to 54km h^{-1}). Yet, most wind turbines operate in a range of wind speeds from around 4m s^{-1} to around 25m s^{-1}.

- The capacity factor of renewable energy does not tell us anything about 'backup' requirements. The capacity factor simply provides an indication of the amount of energy, on average, that a given capacity of renewable plant would be expected to provide per year. As described in the section on 'Power system reliability and operation', the scale of actions needed to manage intermittency is derived statistically.
- However, capacity factor does indicate the theoretical maximum size of the comparator plant against which intermittent generators should be assessed when determining what is required to maintain reliability. A 1000 megawatt (MW) wind farm with a 30 per cent capacity factor delivers the same energy as a 350MW modern gas power station, allowing for the 15 per cent outage rate typical of such generators. Hence, in energy terms, the maximum amount of conventional generation that such a wind farm would displace is 350MW. Even if the wind farm cannot contribute anything to reliability, its 'backup' requirement cannot exceed the amount of conventional generation it displaces (i.e. 350MW/1000MW = 35 per cent of its installed capacity). This is why claims that renewable generators need 100 per cent (or even 60 or 70 per cent) 'backup' per megawatt installed are muddled and incorrect.
- The latter assertion conflates energy and power. Intermittent sources are unlikely to be able to provide the same level of reliable *power* output during demand peaks as a conventional generator. This *will usually* give rise to a need for additional capacity to maintain reliability, particularly at larger penetrations of intermittent sources. However, CO_2 reductions are a function of the *total energy* provided by intermittent stations and, hence, fossil fuel use avoided, not output at peak demand periods.
- Confusion arises because the share of total *energy* provided by an intermittent station may be larger than its contribution to reliability. In fact, even if the contribution of an intermittent source at peak periods is expected to be zero (as would be the case for PV power in the UK, for example), its contribution to CO_2 savings are still a direct function of its energy output.
- Actual CO_2 savings are dependent upon what fossil fuel plant is displaced. These savings are reduced by efficiency losses in thermal plant affected by intermittency and additional use of reserve and response services. In practice, these losses are a small proportion of the energy provided. CO_2 savings are, within a few percentage points, directly linked to the energy that renewable stations generate (Gross et al, 2006).

QUANTITATIVE FINDINGS ON IMPACTS AND COSTS

History of research on intermittency

Early studies: 1970s and 1980s
Many of the studies from the late 1970s to the late 1980s were carried out by, or for, what were then state-owned utilities and in response to the Organization of the Petroleum Exporting Countries (OPEC)-induced oil price shocks,

with a large number focusing on the basic principles of how to represent intermittent generators on an integrated network. Most are concerned with transmission system-level reliability, reserve and supply–demand balancing issues, with a particular emphasis on the role of wind and other renewables as 'fuel savers'. In all cases, the context was very different from that of today: the centralized operation of electricity networks was then still in existence in all countries. As a result, optimization of networks with intermittent sources was conceptualized in rather different terms than it is currently, although the technical issues are largely unchanged. Many reports are concerned with the development of methodological principles and apply these only to relatively simple – and obviously at that time hypothetical – scenarios of renewables penetration (a full list of references to these early studies, and to more recent work, is included in Gross et al, 2006).

Methodological refinement: The 1990s

During the 1990s, there was a marked decrease in the number of utility studies compared to the early 1980s, although academic work continued in the US, UK and Nordic countries. One notable addition to the body of knowledge during this period was a series of ten country studies sponsored by the European Commission. In addition, the breakup of national monopolies is possibly reflected in a marked reduction in emphasis on the *benefits* of wind and other renewables (e.g. fuel saving and system optimization). Instead, work during this period has a noticeable focus on detailed methodological issues and, in particular, on *costs* of system balancing and calculation of capacity margin and other measures of system reliability. Several studies pay attention to methodological refinement and development through, for example, incorporation of stochastic variables into simulation models.

Recent research

The beginning of the 21st century saw a very significant increase in research activity on intermittency. More than 70 per cent of the references in the UKERC database date from the year 2000 onwards. While most utilities have been privatized, system operators, regulators and governments have funded a significant number of studies. Attention to methodological issues has been sustained and extended. In addition, an increasing amount of empirical data has been combined with increasingly sophisticated scenarios of wind power and other intermittent generation installation.

General observations on findings

In what follows, we provide a brief review of UKERC's findings related to additional system balancing requirements and the associated costs, and the range of capacity credits attributed to intermittent sources. We consider why controversy can arise when considering the costs of maintaining reliability, and present UKERC's approach to reconciling this.

The UKERC report also considers a range of other system balancing impacts, such as energy spilling and efficiency losses for market participants.

In general, these impacts are shown to be small and are not discussed in this chapter.

Nearly all the studies reviewed focused exclusively on wind generation, reflecting the relatively advanced penetration of wind power.

Findings for additional system balancing requirements

The principal range of findings of additional reserve and response requirements attributable to intermittency are presented in Figure 4.1. Note that this figure only presents findings that used a common *metric* for the measures of system balancing requirements. Studies that use different approaches are documented separately in Gross et al (2006).

The high outliers are from a German study (E.ON Netz, 2004). It is not clear whether the E.ON Netz reserve requirements refer to balancing services only or also include an element of capacity provision that reflects the relatively low capacity credit of German wind farms.[11] Moreover, particular difficulties are faced within the E.ON Netz region, which has extensive wind energy developments:

- factors that tend to exacerbate the scale of swings in output: low average wind speeds and, thus, low capacity factor for wind output, and substantial 'clustering' of wind farms in the north-west of the control area;
- limited interconnection with regions to the east and west;

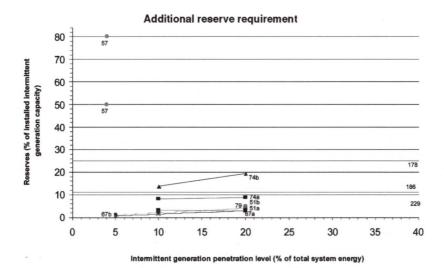

Source: 51: Mott MacDonald (2003); 57: E.ON Netz (2004); 67: Holttinen (2004); 74: DENA Project Steering Group (2005); 79: Dale et al (2003); 178: Doherty et al (2004); 186: Milligan (2001); 229: Hudson et al (2001). The code number assigned to each source refers to entries in the database built up by the UKERC team during the course of the study

Figure 4.1 *Range of findings related to additional reserve requirement with increasing penetration of intermittent supplies*[10]

- 'gate closure' 24 hours ahead of real time – compared to 1 hour in the UK – which means that the forecasting error that must be managed by reserve plants is much larger than in countries with shorter periods between scheduled generation unit commitment and real time.

It is worth noting that different analysts use different definitions of 'reserves', which means that a range of impacts are captured. For example, some studies look exclusively at spinning reserve (part-loaded plant) and so have not included the impact of intermittent generation on other system balancing services, such as the level of standing reserve. Others identify figures for frequency control and load following reserve, but do not analyse the impact on generating unit commitment (the requirement to instruct plant in advance of when it is required to allow sufficient time for it to be brought into operation). These factors can account for some of the differences in findings between studies.

Findings for additional system balancing costs

Twenty-three studies provide quantitative evidence on costs associated with additional reserve and response requirements attributable to the addition of intermittency. The main findings are represented in Figure 4.2. The shaded area on Figure 4.2 represents the range of values taken from UK studies.

There are three very high outliers on Figure 4.2. The first, UK£8.1 per megawatt hour (MWh) (Milborrow, 2005), presents data from the E.ON Netz 2004 report, which lead to high estimates for the reasons discussed with reference to Figure 4.1. The other two high outliers of UK£5.6/MWh and £8.4/MWh are both from a Danish study that appears to amalgamate balancing and system reliability costs (Bach, 2004). By contrast, a Danish study focused on balancing cost alone provides a much lower estimate of cost (Pedersen et al, 2002). The other relatively high figure of UK£4.3/MWh to £4.8/MWh (Fabbri et al, 2005) is from a study that considers the *price* of individual wind farm operators procuring electricity in the Spanish balancing market to cover their positions. This reflects the difference between predicted and actual generation from individual wind plant, rather than the *cost* of system-level balancing.

All the studies that show a range of penetration levels find that reserve costs tend to rise as penetration level increases; but the range of costs across studies is broadly similar at each penetration level (i.e. there is no appreciable convergence or divergence as penetration rises). The difference *between* individual studies is typically larger than the increase in costs *within* each study resulting from increasing penetration levels, which suggests that the reserve cost is particularly sensitive to assumptions about system characteristics, existing reserves, and what is included within the definition of reserve requirements.

Findings for additional capacity requirements to ensure reliability

Twenty-nine studies provide quantitative evidence on the capacity credit of intermittent generators. All use a statistical or simulation approach based upon

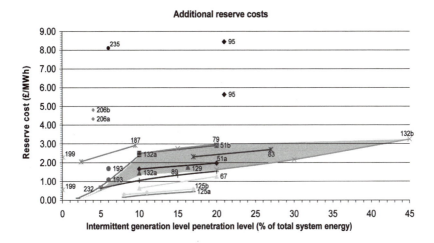

Source: 51: Mott MacDonald (2003); 67: Holttinen (2004); 79: Dale et al (2003); 83: ILEX and Strbac (2002); 89: Milborrow (2004); 95: Bach (2004); 125 ILEX et al (2004); 129: Pedersen et al (2002); 132: Milborrow (2001); 187: Seck (2003); 193: Hirst (2002); 199: Hirst (2001); 206: Fabbri et al (2005); 232: Dale (2002); 235: Milborrow (2005). The code number assigned to each source refers to entries in the database built up by the UKERC team during the course of the study

Figure 4.2 *Range of findings on the cost of additional reserve requirements*[12]

a measure of reliability such as LOLP. The main findings are represented in Figure 4.3. The shaded area on Figure 4.3 covers the range of values taken from UK studies.

The findings demonstrate the sensitivity of the capacity credit to resource availability and the degree of correlation between resource availability and periods of high demand. Capacity credit values are adversely affected where there is a low degree of correlation between resource availability and peak loads. Studies relevant to British conditions, all of which focus on wind power, indicate that output and demand are largely uncorrelated on a diurnal basis (even though wind speeds *are* generally higher in winter).

The relationship between resource and capacity credit is also demonstrated by studies using data from operating wind farms in a region with low average wind speeds. For example, two German studies (DENA Project Steering Group, 2005; E.ON Netz, 2005) show the effect that the relatively weak wind resource in Germany has on the capacity credit value.

Additional capacity costs

System balancing services are purchased by the transmission system operator through advance contracts and are also provided through actions taken by the system operator in the balancing market. As such, the costs of these impacts are readily quantified and not subject to much controversy.

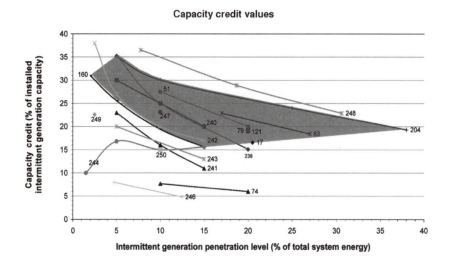

Figure 4.3 *Range of findings on capacity credit of intermittent generation*[13]

Source: 17: Watson (2001); 51: Mott MacDonald (2003); 74: DENA Project Steering Group (2005); 79: Dale et al (2003); 83: ILEX and Strbac (2002); 121: Giebel (2000); 160: Holt et al (1990); 204: Grubb (1991); 238: Martin and Carlin (1983); 240: Commission of the European Union (1992b); 241: Danish Energy Ministry (1983); 242: Commission of the European Union (1992d); 243: Commission of the European Union (1992a); 244: Commission of the European Union (1992g); 246: E.ON Netz (2005); 247: Sinden (2005); 248: Commission of the European Union (1992f); 249: Commission of the European Union (1992e); 250: Commission of the European Union (1992c). The code number assigned to each source refers to entries in the database built up by the UKERC team during the course of the study

Reliability costs are less straightforward. In many cases, adding intermittent generators to an electricity network will tend to increase the total amount of plant required on the system to provide a given measure of reliability relative to delivering the same energy with all thermal plant. This is because the capacity credit of intermittent generation tends to be smaller than the contribution to reliability of thermal generation that delivers the same energy output. However, in Britain at least, no direct payment is made for 'reliability reserves'. System margin emerges from an aggregated set of investment decisions made by market participants. Unlike the additional reserve and response services that intermittency might give rise to, the system operator does not contract for plant in order to maintain system margin or to act as 'backup' to intermittent generators. How, then, can one calculate the costs of maintaining reliable supplies?

Two distinct lines of thought may be found in the literature. The first is concerned only with the *total change* in costs for a 'with-intermittent' system relative to a 'no-intermittent' comparator (Dale et al, 2003). This is calculated by comparing a system that contains intermittent generators with one that meets the same reliability criteria without those intermittent generators –

assuming that both systems have the same energy output. More plant is required than would be the case in the absence of intermittent stations since the contribution of the intermittent generators to reliability (such as LOLP; see the earlier section on 'Power system reliability and operation') is lower than that of thermal generators. However, this approach does not attempt to directly attribute a cost of capacity reserves due to intermittent stations (Milborrow, 2001; Dale et al, 2003). The reason for this is that there is no explicit market for, or central procurer of, such services. The main advantage of this approach is that it is perfectly consistent with current market arrangements. Its principal disadvantage is that it does not readily permit a like-with-like comparison between the generating costs of different types of generator (e.g. wind versus coal) that includes an explicit cost of intermittency.[14]

A second line of thought directly costs the additional 'capacity reserve' put in place to ensure reliability (ILEX and Strbac, 2002). 'Backup' or 'capacity reserve' sufficient to close any gap between the capacity credit of intermittent stations and that of conventional generation which would provide the same amount of energy is explicitly costed. The principal problem with this approach is that it depends upon an assumption about what form of generation is to provide the backup. Costs will vary depending upon the nature of this assumption. It is also not clear that we can know the long-run marginal cost of such capacity as this will be a product of future system optimization (market based or otherwise), which will be affected by new technologies or practices.

A simple algebraic exposition was developed for the UKERC report that allows both techniques to be reconciled.[15] An identity can be derived for estimating the capacity credit-related cost of intermittency. This shows that the system reliability cost of intermittency = fixed cost of energy-equivalent thermal plant[16] minus fixed cost of thermal plant displaced by capacity credit of the intermittent plant.

This approach allows the capacity credit-related costs associated with adding intermittent plant to the system to be made explicit in a way that is consistent with systemic principles, making no judgement about the nature of the plant that *actually* provides capacity to maintain reliability. All that is required is determination of the *least cost energy equivalent comparator* (i.e. the thermal plant that would supply the same energy in the absence of intermittent generation).

Table 4.1 takes a range of capacity credits for 10 and 20 per cent penetration of wind energy based upon the range of UK-relevant findings in Figure 4.3. We combined this range of capacity credits with fixed data for total system size, thermal-equivalent capacity costs, thermal-equivalent capacity factor, and wind capacity factor. These data represent a future Great Britain electricity system with a least-cost thermal generation comparator and wind generation.[17] In each illustration, the only figures changed are the capacity credit and total energy contribution from wind.[18]

If capacity credit were zero (which would imply that there was a zero probability of wind power being available during periods of peak demand), and all other characteristics held as per Table 4.1, the costs of maintaining reliability would rise to UK£9/MWh of wind energy.

Table 4.1 *Relationship between capacity credit and reliability cost, Great Britain-relevant capacity credits and system characteristics*

Wind energy penetration level	Capacity credit range	Reliability cost (UK£/MWh of wind)
10% (40TWh of wind energy; 13GW of wind installed)	19.4%	£4.76
	30.0%	£2.44
20% (80TWh of wind energy; 26.1GW of wind installed)	19.1%	£4.82
	26.0%	£3.32
System characteristics[19]		
Total system energy	400TWh yr^{-1}	
Wind capacity factor	35%	
Thermal-equivalent capacity factor[20]	85%	
Thermal-equivalent capacity cost	£67,000/MW/year	

Source: Gross et al (2006)

SUMMARY OF FINDINGS AND CONCLUSIONS

General comments

Unless the assumptions and characteristics of the system being analysed are very clearly understood, there is a danger that the results will be misinterpreted or that invalid comparisons will be drawn. It is apparent from an analysis of each study that the results of any individual work are sensitive to a set of system characteristics:

- the existing generation mix (in particular, the degree of flexibility of existing plant and suitability for part loading, and the rate at which existing plant can increase or decrease output);
- existing requirements for reserve services for system balancing;
- the spatial distribution of intermittent generation plant;
- the mix of intermittent generation technologies;
- transmission network constraints and the size of links to other networks;
- the absolute level of renewable resource available and the degree of correlation of resource availability with demand peaks and troughs;
- generating unit commitment time horizon and accuracy of renewable resource forecasting;
- the overall system reliability/security target level.

It is important to note that data limitations, methodological details and the scope of impacts/costs may differ between studies. It is only possible for this chapter to highlight significant outliers and general trends.

Relevance of simulation and empirical studies

The majority of the studies reviewed use simulated data, real data extrapolated or real data run through a range of models. The main exception is experience from Germany's E.ON Netz, which tends to show relatively high costs for reserves. Moreover, it has been contended that experience in Denmark and Germany suggests that simulation studies in the UK may have failed to capture the extent of prospective fluctuations.[21] However, it is also important to note that experience cannot supersede simulation if the experience is not directly relevant. We would not conclude, for example, that PV should have a significant capacity credit in the UK because of experience with solar plants in California. It is also important to note that there are important differences between Denmark, Germany and the UK:

- Denmark is a small country and the scope for geographical dispersion is limited. The system must also integrate output from heat demand-constrained combined heat and power (CHP) plant, and has a very high penetration level of wind energy (Pedersen et al, 2002; Bach, 2004; Holttinen, 2004).
- Denmark is heavily interconnected to both the Nordel and German electricity systems and, hence, is able to manage intermittency in ways that are unavailable to the UK.
- We have discussed some of the differences between Britain and Germany (most notably, the lower capacity factor of German wind farms), and the specific issues that relate to the geography and operating practices of the E.ON Netz region. It is also clear that the DENA grid study (2005), which looks at a wider geographical area, takes a more optimistic view than E.ON Netz (2005).

It is important that key problems are not 'assumed away'. Some existing studies explicitly explore key effects, such as regional concentration of some renewables (ILEX and Strbac, 2002). However, others have assumed that wind energy will be geographically dispersed and, hence, may have failed to identify an important prospective cost. It has been suggested that wind developments tend to cluster in areas with good wind resources and that, in the future, large individual offshore developments may present problems for system operators.[22] These impacts must be explored in analytic research and monitored as empirical evidence increases.

The estimates presented below assume that intermittent generation is primarily wind, that it is geographically widespread, and that it accounts for no more than about 20 per cent of electricity supply. At current penetration levels, costs are much lower since the costs of intermittency rise as penetrations increase. If intermittent generation were clustered geographically, or if the market share were to rise above 20 per cent, intermittency costs would rise above these estimates and/or more radical changes would be needed in order to accommodate renewables.

System balancing requirements

The majority of the studies that are applicable to the UK find that, up to an intermittent generation penetration level of 20 per cent, the additional reserve requirements imposed on the system are generally less than 10 per cent of the installed capacity of the intermittent generators. The studies that present higher reserve requirements either represent systems which are not directly comparable to the UK or use a methodological approach that is not consistent with widely accepted practice.

Over 80 per cent of the studies concluded that the cost of providing additional reserves would be less (and, in many cases, substantially less) than UK£5/MWh of intermittent generation at intermittent generation penetration levels up to, and in some cases exceeding, 20 per cent. Studies with assumptions relevant to the British electricity system fall into the range of UK£2/MWh to £3/MWh. Those remaining studies that present higher costs relate either to systems with much higher penetration levels, or to where resource availability is not comparable with Britain, or are based on methodology that is inconsistent with UK regulatory and system operation practices.

System reliability requirements

Capacity credit is a measure of the contribution that intermittent generation can make to reliability. It is usually expressed as a percentage of the installed capacity of the intermittent generators. There is a range of estimates for capacity credits in the literature and the reasons for there being a range are well understood. The range of findings relevant to British conditions is approximately 20 to 30 per cent of installed capacity when up to 20 per cent of electricity is sourced from intermittent supplies (usually assumed to be wind power). Capacity credit as a percentage of installed intermittent capacity declines as the share of electricity supplied by intermittent sources increases.

Using the capacity credit values applicable to the British electricity system and weather conditions, and the formula described earlier (system reliability cost of intermittency = fixed cost of energy-equivalent thermal plant minus fixed cost of thermal plant displaced by capacity credit of the intermittent plant), the costs of maintaining reliability fall into the range of UK£3/MWh to £5/MWh for penetrations of up to 20 per cent.

Total costs

The total additional cost that intermittency imposes is the sum of system balancing costs plus the costs of maintaining reliability. The weight of evidence suggests that in Britain, these costs are likely to lie in the range of UK£5/MWh to £8/MWh (0.5 pence/kWh to 0.8 pence/kWh) of intermittent output. If shared between all electricity consumers, the impact on electricity prices would be of the order of 0.1 pence/kWh to 0.15 pence/kWh, at a time when retail electricity prices are around 9 pence/kWh.

Implications and recommendations for policy and further research

It is important that policy-makers and regulators continuously monitor the effect of intermittent generation on system margin, the nature of other new plant, and the location and concentration of renewables development. To keep costs low, policy needs to encourage:

- wide geographical dispersion of renewables;
- a diversity of renewables;
- investment in flexible and reliable generation and more demand-response capability;
- ongoing monitoring of measures of reliability;
- ongoing research is also needed into a range of issues related to such things as the impact of regional clustering, measures of reliability, the need for policy to ensure reliability, and long-term options for managing intermittent output.

NOTES

1   Terminology differs between authors, and many analysts advocate the use of 'variable' in preference to 'intermittent', noting that all power sources are interruptible and, hence, intermittent. UKERC acknowledged the semantic difficulties and chose a pragmatic course of adopting what is at the time of writing the most widely used, albeit potentially inaccurate, term.
2   Supply-side adjustments are the current norm because most consumers are not under the control of system operators or direct participants in wholesale electricity markets. Some very large consumers (such as steel works or chemical plants) can participate in wholesale markets and reduce demands in response to high prices, or contract to respond to requests from the system operator.
3   System margin is the current UK Grid Code term. The concept has been referred to historically as 'capacity margin', 'system reserves' and 'plant margin'.
4   See National Grid plc (2005/2006) *Winter Outlook Report*, www.ofgem.gov.uk/temp/ofgem/cache/cmsattach/12493_214_05.pdf?wtfrom=/ofgem/index.jsp.
5   National Grid anticipated that excess generation over average weekly peak winter demand is 12GW to 14GW; see National Grid plc (2005/2006) *Winter Outlook Report*, www.ofgem.gov.uk/temp/ofgem/cache/cmsattach/12493_214_05.pdf?wtfrom=/ofgem/index.jsp.
6   Standard deviation is a measure of how tightly clustered a set of values are around the mean value of the set of data (i.e. it is a measure of the spread of the data).
7   Variance is the square of the standard deviation.
8   Some commentators have noted that although the risk of capacity shortages is highest at times of peak demand, it may not be much lower within a few gigawatts

of peak because the standard deviation of available thermal capacity at peak can be nearly 2GW.

9   For example, 'wind turbines are completely effete because they need backup all the time and help to produce CO_2, not reduce it' (David Bellamy, BBC Radio 4 'Today' programme, 18 November 2005); see also www.countryguardian.net/ for further examples.

10 In this figure we present findings that estimate additions to reserves in two ways, or *metrics*:

1   as a percentage of installed intermittent generation capacity at given levels of intermittent generation penetration, and where penetration level is expressed as the percentage of *total system energy* provided from intermittent generation (these appear as a point or series of points in Figure 4.1);

2   as a percentage of installed intermittent generation capacity, but *no penetration level given* (these appear as a horizontal line in Figure 4.1).

11 E.ON Netz introduce the term 'shadow capacity', which is not used in any other literature and its precise meaning is unclear.

12 Costs have been converted to UK sterling using exchange rates at the date of publication and values inflated to 2005 using producer price index. All values are per megawatt hour of intermittent output. In this figure we present findings that used the following approach: cost per megawatt hour of electricity from intermittent generation at given levels of intermittent generation penetration, where penetration level is expressed as the percentage of *total system energy* provided from intermittent generation. Fifteen studies used this approach.

13 In this figure we present findings where capacity credit is expressed as a percentage of installed intermittent generation capacity at given levels of penetration, with penetration level expressed as the percentage of total system energy provided from intermittent generation. Nineteen studies used this approach.

14 It is possible to derive the cost of maintaining reliability using the above approach by assessing the impact on system load factors (Dale et al, 2003). This is because one effect of adding intermittent generators is that the load factor of the remaining conventional stations on the system will fall since additional capacity is needed to provide a given energy output.

15 See UKERC working paper 'Methods for reporting costs related to the capacity credit of intermittent generation relative to conventional generation' at www.ukerc.ac.uk/content/view/332/233.

16 This is the capacity of thermal plant, operated at maximum load factor, which would provide the same amount of energy as the intermittent stations under investigation.

17 The data are derived from a recent and widely cited UK study (Dale et al, 2003).

18 This is a simplification since capacity factor and capacity credit are related variables. Nevertheless, the range of capacity credits in Table 4.1 is consistent with the capacity factors and other characteristics reported in the table. It is illustrative of the likely range relevant to UK conditions and assumes that the predominant intermittent source is wind power.

19 Assumptions taken from Dale et al (2003) and seeking to represent a future Great Britain electricity system with demand of 400TWh yr^{-1}, a mix of onshore and offshore wind, and where combined-cycle gas turbine (CCGT) continues to provide the least-cost form of new electricity generation plant.

20 A key principle of this approach is that comparator plant is assumed to be lowest-cost new generation. Such plant would be operated at maximum capacity factor (CF) and is assumed here to be CCGT. We use 85 per cent CF as an approxima-

tion; in fact, some new plant exceeds this. Availability at peak demand is probably higher (above 90 per cent; see National Grid plc (2005/2006) *Winter Outlook Report*, www.ofgem.gov.uk/temp/ofgem/cache/cmsattach/12493_214_05.pdf?wt from=/ofgem/index.jsp), while system load factor (typically around 58 per cent) or the load factor of the entire fleet of CCGT, as operated at present (typically around 60 per cent and affected by gas prices and other market factors), are both lower. The methodology is predicated on a 'like-with-like' comparison between a new thermal station and intermittent plant, both of which operate at maximum output.

21 Hugh Sharman, presentation to the UKERC Stakeholder Workshop, available at www.ukerc.ac.uk/content/view/332/233.

22 Hugh Sharman, presentation to the UKERC Stakeholder Workshop, available at www.ukerc.ac.uk/content/view/332/233, and personal communication with Hannele Holttinen, 2006.

REFERENCES

Bach, P. (2004) *Costs of Wind Power Integration into Electricity Grids: Integration of Wind Power into Electricity Grids Economic and Reliability Impacts*, IEA Workshop on Wind Integration, Paris, www.iea.org/textbase/work/2004/nea/ bach.pdf

Commission of the European Union (1992a) *Wind Power Penetration Study: The Case of the Netherlands*, CEC ref EUR 14246 EN, CEC, Brussels, Luxembourg, p46

Commission of the European Union (1992b) *Wind Power Penetration Study: The Case of Denmark*, CEC ref EUR 14248 EN, CEC, Brussels, Luxembourg, p124

Commission of the European Union (1992c) *Wind Power Penetration Study: The Case of Germany*, CEC ref EUR 14249 EN, CEC, Brussels, Luxembourg, p68

Commission of the European Union (1992d) *Wind Power Penetration Study: The Case of Greece*, CEC ref EUR 14252 EN CEC, Brussels, Luxembourg, p71

Commission of the European Union (1992e) *Wind Power Penetration Study: The Case of Italy*, CEC ref EUR 14244 EN, CEC, Brussels, Luxembourg, p73

Commission of the European Union (1992f) *Wind Power Penetration Study: The Case of Portugal*, CEC ref EUR 14247 EN, CEC, Brussels, Luxembourg, p104

Commission of the European Union (1992g) *Wind Power Penetration Study: The Case of Spain*, CEC ref EUR 14251 EN, CEC, Brussels, Luxembourg, p81

Dale, L. (2002) *Neta and Wind*, EPSRC Blowing Workshop, UMIST, Manchester, www.ee.qub.ac/blowing/activity/UMIST/WS3_Lewis_Dale.pdf

Dale, L., Milborrow, D., Slark, R. and Strbac, G. (2003) 'A shift to wind is not unfeasible (total cost estimates for large-scale wind scenarios in UK)', *Power UK*, issue 109, pp17–25

Danish Energy Ministry (1983) *Vindkraft i Elsystemet* (Wind power in the electricity system), Danish Energy Ministry, Copenhagen, EEV 83-02

DENA Project Steering Group (2005) *Planning of the Grid Integration of Wind Energy in Germany Onshore and Offshore up to the Year 2020*, DENA grid study, Deutsche Energie-Agentur, Berlin, www.wind-energie.de/fileadmin/dokumente/ Themen_A-Z/Netzausbau/stud_summary-dena_grid.pdf

Doherty, R., Bryans, L., Gardner, P. and O'Malley, M. (2004) 'Wind penetration studies on the island of Ireland', *Wind Engineering*, vol 28, issue 1, pp27–42

E.ON Netz (2004) *Wind Report 2004*, E.ON Netz GMBH, Bayreuth, Germany

E.ON Netz (2005) *Wind Report 2005*, E.ON Netz GMBH, Bayreuth, Germany

Fabbri, A., Gomez San Roman, T., Rivier Abbad, J. and Mendez Quezada, V. H. (2005) 'Assessment of the cost associated with wind generation prediction errors in a liberalized electricity market', *IEEE Transactions on Power Systems*, vol 20, issue 3, pp1440–1446

Giebel, G. (2000) *The Capacity Credit of Wind Energy in Europe, Estimated from Reanalysis Data*, Risø National Laboratory Working Paper, Conference proceeding, Global Dialogue EXPO 2000, Hanover, 10–17 July

Gross, R., Heptonstall, P., Anderson, D., Green, T., Leach, M. and Skea, J. (2006) *The Costs and Impacts of Intermittency: An Assessment of the Evidence on the Costs and Impacts of Intermittent Generation on the British Electricity Network*, UK Energy Research Centre, London

Grubb, M. J. (1991) 'The integration of renewable electricity sources', *Energy Policy*, vol 19, issue 7, pp670–688

Hirst, E. (2001) *Interactions of Wind Farms with Bulk-Power Operations and Markets*, Project for Sustainable FERC Energy Policy, Virginia, US, http://cwec.ucdavis.edu/rpsintegration/library/Wind%20farms%20and%20bulk-power%20interactions%20Sep01%20Hirst.pdf

Hirst, E. (2002) *Integrating Wind Energy with the BPA Power System: Preliminary Study*, Prepared for Power Business Line, Bonneville Power Administration, Oregon, US, by Consulting in Electric-Industry Restructuring, Oak Ridge, Tennessee, US, www.bpa.gov/power/pgc/wind/Wind_ Integration_Study_09-2002.pdf

Holt, J. S., Milborrow, D. and Thorpe, A. (1990) *Assessment of the Impact of Wind Energy on the CEGB System*, CEC contract ref EN3W. 0058. UK (H)CEC, Brussels

Holttinen, H. (2004) *The Impact of Large Scale Wind Power Production on the Nordic Electricity System*, PhD thesis, Helsinki University of Technology, Department of Engineering, Physics and Mathematics, Helsinki, Finland

Hudson, R., Kirby, B. and Wan, Y. (2001) *The Impact of Wind Generation on System Regulation Requirements*, American Wind Energy Association, Windpower 2001 Conference Proceedings, Washington, DC

ILEX and Strbac, G. (2002) *Quantifying the System Costs of Additional Renewables in 2020*, ILEX for DTI, London, www.dti.gov.uk/energy/developep/080scar_report_v2_0.pdf

ILEX; The Electricity Research Centre, University College Dublin; The Electric Power and Energy Systems Research Group, The Queen's University Belfast; Manchester Centre for Electrical Energy, The University of Manchester Institute of Science and Technology (2004) *Operating Reserve Requirements as Wind Power Penetration Increases in the Irish Electricity System*, Sustainable Energy Ireland, www.sei.ie/uploadedfiles/InfoCentre/Ilex

Martin, B. and Carlin, J. (1983) 'Wind-load correlation and estimates of the capacity credit of wind power: An empirical investigation', *Wind Engineering*, vol 7, issue 2, pp79–84

Milborrow, D. (2001) *Penalties for Intermittent Sources of Energy*, Working Paper for PIU Energy Review, www.pm.gov.uk/files/pdf/Milborrow.pdf

Milborrow, D. (2004), *Perspectives from Abroad: Assimilation of Wind Energy into the Irish Electricity Network*, Sustainable Energy Ireland, Dublin

Milborrow, D. (2005) 'German report skews picture of wind on the grid', *Windstats Newsletter*, vol 18, issue 1, pp1–2

Milligan, M. (2001) *A Chronological Reliability Model to Assess Operating Reserve Allocation to Wind Power Plants*, National Renewable Energy Laboratory, 2001 European Wind Energy Conference, www.nrel.gov/docs/fy01losti/30490.pdf

Mott MacDonald (2003) *The Carbon Trust and DTI Renewables Network Impact Study Annex 4: Intermittency Literature Survey and Roadmap*, The Carbon Trust and DTI, London, www.thecarbontrust.co.uk/carbontrust/about/publications/Annex4.pdf

Pedersen, J., Eriksen, P. B. and Mortensen, P. (2002) *Present and Future Integration of Large-Scale Wind Power into Eltra's Power System*, Eltra, Denmark, www.eltra.dk/media/showMedium.asp?14556_LCID1033

Seck, T. (2003) *GRE Wind Integration Study*, Great River Energy, UWIG Technical Wind Workshop Proceedings, www.uwig.org/seattlefiles/seck.pdf

Sinden. G. (2005) *Wind Power and the UK Resource*, Environmental Change Institute, University of Oxford, for DTI, London, www.eci.ox.ac.uk/renewables/UKWind-Report.pdf

Sorrell, S. (2007) 'Improving the evidence base for energy policy: The role of systematic reviews', *Energy Policy*, vol 35, pp1858–1871

Watson, R. (2001) 'Large scale integration of wind power in an island utility: An assessment of the likely variability of wind power production in Ireland', in *IEEE Power Tech Conference Proceedings*, Porto, Portugal, vol 4, p6

5

Wind Power Forecasting

Bernhard Lange, Kurt Rohrig, Florian Schlögl,
Ümit Cali and Rene Jursa

INTRODUCTION

Electricity generated from wind power will play an important role in future energy supply in many countries. This implies the need to integrate this power within the existing electricity supply system, which was mainly designed for large units of fossil fuel and nuclear power stations. Wind power has different characteristics; therefore, this integration leads to some important challenges from the viewpoint of the electricity system.

The availability of power supply generated from wind energy varies fundamentally from that generated conventionally from fossil fuels. The most important difference is that wind power generation depends upon the availability of wind (i.e. it is weather dependent). In contrast to conventional power plants, which are controlled to produce power according to demand, wind power is usually produced according to the available wind. This also means that the power output fluctuates with wind variations. In the electricity system, supply and demand must be equal at all times. Thus, in an electricity system with an important share of wind power, new methods of balancing supply and demand are needed.

Wind power forecasting plays a key role in tackling this challenge. It is the prerequisite for integrating a large share of wind power in an electricity system since it links weather-dependent production with the scheduled production of conventional power plants and forecasts of electricity demand, the latter being predictable with reasonable accuracy. This is illustrated in the following example.

Figure 5.1 shows the electricity demand in Germany for one week as an example (upper curve). The first day was a public holiday, with a relatively low electricity demand, compared to that of day four, which was a Sunday. Saturday shows a slightly higher demand and also a different temporal course. The load was again slightly higher on the Friday between the holiday and the weekend. The last three days show the typical weekday load curves. This load curve is readily predictable, even for a rather atypical week, as shown in the example, and the conventional power plants are scheduled such that their

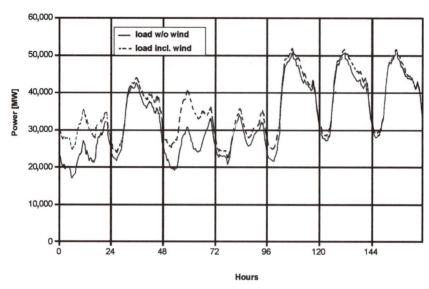

Figure 5.1 *Load and wind power generation for one week in Germany (2003)*

production follows the predicted load curve. Deviations of the actual from the predicted load are equalized by using balancing power.

The dark band shown in Figure 5.1 was the share of electricity generated by wind power in Germany during this week. The wind power production varied between almost zero on the last day and up to about 10 gigawatts (GW) on day three. Conventional power plants had to supply only the share of the load shown by the lower curve. If wind power generation was not predicted, it would appear as an additional and unknown 'negative load' and would require an extremely large use of balancing energy. This is technically and economically undesirable. Instead, the forecast power output from wind power is used together with the load forecast to schedule the conventional power plants. In this way, only the errors in the forecasts have to be balanced by balancing energy. This also clearly shows that the forecast error determines the need of balancing energy in order to integrate wind power.

A wind power forecast is indispensable for system operation and security. Its accuracy is directly connected to the need for balancing energy and, hence, to the cost of wind power integration. Consequently, a large amount of research has been directed towards developing high-quality and reliable wind power forecasts during recent years, and many different forecasting systems with different approaches have been developed. In countries with a substantial share of wind power in the electricity system, such as Denmark, Germany or Spain, wind power forecasting systems are already an essential part of grid and system control.

APPLICATIONS OF WIND POWER FORECASTING

The most important application for wind power forecasting is to reduce the need for balancing energy and reserve power, which are needed to integrate wind power within the balancing of supply and demand in the electricity supply system (i.e. to optimize power plant scheduling). This leads to lower integration costs for wind power, lower emissions from the power plants used for balancing, and, subsequently, to a higher value of wind power.

A second application is to provide forecasts of wind power feed-in for grid operation and grid security evaluation. To assess the security of the grid and to operate it (e.g. for maintenance and repair), the grid operator needs to know the current and future wind power feed-in at each grid connection point.

The objectives of a wind power forecast therefore depend upon the application:

- For optimized power plant scheduling and power balancing, an accurate forecast of the wind power generation for the whole control zone is needed. The relevant time horizon depends upon the technical and regulatory framework (e.g. the types of conventional power plants in the system and the trading closure times).
- For determining the reserve power that has to be held ready to provide balancing energy, a prediction of the accuracy of the forecast is needed. Since the largest forecast errors determine the need for reserve power, these have to be minimized. In Germany, the relevant forecast horizon is usually rather long (i.e. predicted one day ahead).
- For grid operation, the current and forecast wind power generation in each grid area or grid connection point is needed. This requires a forecast for small regions or even single wind farms. For grid management, shorter time horizons are often relevant. Switching and other grid operations do not have a long lead time and therefore a higher accuracy of short-term forecasts is more important.

STEPS IN A FORECASTING SYSTEM

In producing a wind power forecast, different steps can be distinguished:

- numerical weather prediction;
- wind power output forecast;
- regional upscaling.

As the first step, a weather prediction, including the forecast of the wind speed and possibly some other meteorological parameters, is needed for a wind power forecast. This is provided by numerical weather prediction (NWP) models. Most often, two or more hierarchical levels with different NWP models and increasing resolution are used (see the section on 'Numerical weather prediction'). Very simple systems use as a substitute for NWP models measured wind speeds from a location in the direction of the mean pressure systems movement. It is possible to compute forecasts without weather prediction from actual measurements of power output, but only for very short forecast horizons.

The NWP data is used as input to the next step: the wind farm power-output forecasting. This takes into account the local meteorological influences on the wind speed and direction, the power conversion characteristics of the turbine, wind farm shading, and other effects that influence the power output. Different approaches and combinations of approaches have been developed and are in use (see the section on 'Different approaches to the power output forecast'). For forecasts with a shorter forecast horizon, online measured wind speeds and/or wind farm power output are used as additional input to the forecasting (see the section on 'Forecast horizon').

If the forecast is needed for a larger region with very many wind farms or wind turbines, forecasts are compiled only for some representative wind farms and the results from these are scaled up to regional forecasts as a third step (see the section on 'Regional upscaling').

NUMERICAL WEATHER PREDICTION

Weather forecasts from numerical weather prediction models (NWP models) are the most essential input needed for almost all wind power forecast models. Usually a model chain of hierarchical levels with different NWP models and increasing resolution is used.

The model chain starts with meteorological measurements and observations all over the globe, performed by meteorologists, weather stations, satellites and so on. All available data are used as input to a global NWP model, which models the atmosphere of the entire Earth. The NWP model calculates the future state of the atmosphere from the physical laws governing the weather. Since these calculations are very computationally expensive, the resolution of a global model has to be rather coarse (see Figure 5.2, left). Global models are in operation at only about 15 weather services.

To provide more accurate weather forecasts, local area models (LAMs) are used, which cover only a small part of the Earth but can be run with a much higher resolution (see Figure 5.2, right). These models use the forecasts of the global model as input and calculate a weather forecast, taking into account the local characteristics of the terrain.

NWP models are usually run operationally by national weather services. Most of these only run a LAM for their region of interest and use data from other global models as input. Some commercial companies also run NWP models, and dedicated service companies also operate NWP models especially for wind power forecasting.

One example of a LAM is the LME model (Doms and Schättler, 1999). It covers central Europe with 325×325 grid cells. This leads to a horizontal resolution of about 7×7km. The forecast horizon of the operational model is 48 hours and the resolution is 1 hour. Model runs are started thrice daily at 00 coordinated universal time (UTC), 12 UTC and 8 UTC.

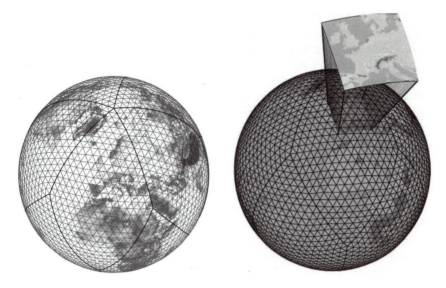

Figure 5.2 *Horizontal grid of a global numerical weather prediction model and enlarged area covered by a local area model*

In some model systems, a third step is performed using a high resolution meso-scale model with an even higher resolution. This is especially important if the LAM available has low resolution and the terrain is complex. The meso-scale models can either be run by the provider) of the weather prediction data (i.e. a weather service provider or as part of the wind power forecast model.

In practical applications, different NWP model data are often available for a wind power forecast. Important criteria for the selection of the most appropriate NWP model(s) are:

- area covered;
- spatial and temporal resolution;
- forecast horizon;
- accuracy;
- number of runs and their calculation time.

DIFFERENT APPROACHES TO THE POWER OUTPUT FORECAST

The aim of a wind power forecast is to link the wind prediction of the NWP model to the power output of the wind farm. Three fundamentally different approaches can be distinguished:

1   The *physical approach* aims to describe the physical process of converting wind to power, and models each of the steps involved.
2   The *statistical approach* aims at describing the connection between predicted wind and power output directly by statistical analysis of time series from data in the past.
3   Finally, the *learning approach* uses artificial intelligence (AI) methods to learn the relationship between predicted wind and power output from time series of the past.

In practical applications, the methods are sometimes combined or mixed. Models with a physical approach almost always use data from the past to tune their models or use model output statistics (MOS) for a correction of the result. On the other hand, models using statistical or AI methods often use knowledge of the physical processes, such as the shape of the power curve, in designing their models.

The physical approach contains a chain of models of the different physical processes involved:

- wind conditions at the site and hub height of the turbines;
- wind farm shading effects;
- turbine power curve;
- model output statistics (MOS).

The wind prediction of a NWP model represents a mean wind speed over the area of one grid cell of the model at a certain height. As a first step, the site-specific wind speed and direction at hub height of the turbines has to be calculated. The models used for this are either micrometeorological models, such as the WAsP model (Mortensen et al, 1993) and/or flow models, usually meso-scale models such as MM5 (Grell et al, 1994). These models take into account the influence of the vertical wind speed profile, the orography of the terrain, the surface roughness and thermal effects. In a second step, wind farm shading effects are calculated by a wind farm model, such as PARK (PARK, 1993) or FLaP (FLaP, 2002). The turbine power curve is then used to convert the wind speed at each turbine into the expected power output. Finally, model output statistics are used to correct the results for systematic deviations caused either by uncertainties in the information needed by the models or by inaccuracies in the models. MOS is a statistical correction method based on time series of the past. It is essential for the good forecasting quality of a physical model since the physical processes are highly complex and the information needed for the models often has limited accuracy. The models require detailed knowledge about the wind farms to be forecast (e.g. the terrain around the wind farm, the layout of the farm and the power curve of the turbines).

Statistical approaches analyse the connection between weather forecasts and power production from time series of the past and describe this connection in a way that enables it to be used for the future.

Like statistical models, artificial intelligence methods also describe the connection between input data (the predictions of the NWP model) and output data (wind farm power output). But instead of an explicit statistical analysis, they use algorithms that are able to implicitly describe non-linear and highly complex relations between this data. Different methods are used for this, such as:

- artificial neural networks;
- support vector machines;
- nearest neighbour search.

For both the statistical and AI approach, long and high-quality past time series of weather predictions and power output are of essential importance.

Each of these approaches is used in practical applications. The physical approach is, for example, used in the Prediktor (Landberg, 1994, 2001) and Previento (Focken et al, 2001; Focken et al, 2002) models. The WPPT (Nielsen and Madsen, 1997; Nielsen et al, 2002) model uses a statistical approach, while the WPMS (Rohrig and Ernst, 2000; Lange et al, 2006) model uses an artificial intelligence approach. For an overview of different models, see Landberg et al (2003), Giebel et al (2003) and Lange and Focken (2005). The main advantages and drawbacks of different approaches for the power output forecasts are summarized in Table 5.1.

Table 5.1 *Summary of the main advantages and drawbacks of different approaches for a power output forecast*

Statistical and artificial intelligence approaches	Physical approach
+ No physical insight necessary	+ Chance to understand physical behaviour
+ Fast calculation	+ Measurement data less important
– Depends upon high-quality and long-term measurement data	– Needs extensive information about wind farms
– Situations with limited numbers of observations are difficult	– Great effort in the set-up

FORECAST HORIZON

The forecast horizon is the time period between the time at which the forecast is available and the forecast point in time. Different forecasts are used for different purposes and their forecast horizons depend upon the requirements of the user, stemming from technical and regulatory conditions, and upon the feasibility of forecasting.

From the meteorological and climatological point of view, one can distinguish long-term or seasonal forecasts with a forecast horizon of several months, medium-term forecasts with a range of up to two weeks, short-term forecasts for the next few days, and very short-term forecasts for a forecast horizon of up to one day. Generally, the forecast accuracy decreases with increasing forecast horizon.

For current wind power forecasting, deterministic forecasts are used up to a forecast horizon of three to five days. Essentially, two forecast horizons have to be distinguished: the day-ahead forecast and the short-term forecast. The day-ahead forecast is mainly used for day-ahead power trading. The forecast horizon therefore depends upon the organization of the trading (e.g. the gate closure time and the trading days). An example for a gate closure time of 12.00 am for the next day is shown in Figure 5.3. The NWP model starts running at midnight with the observations from the day before. It finishes calculation around 7.00 am and sends the information to the wind power forecasting system. This usually has a very short calculation time and the results are available a few minutes later. They are analysed and used for trading the power for the next day until at 12.00 am the trading ends. This means that the calculation of the forecast starts 48 hours ahead, counted from the start of the NWP model. If there is no trading during weekends and pubic holidays, lead time for the calculation for the 'day-ahead' trading can actually be 96 hours or longer.

Short-term wind power forecasting is mainly used for intra-day trading and grid operation and security. Its main characteristic is that it utilizes online data from measurements of actual power output and/or wind speed. For very short forecast horizons, this leads to a very important increase in forecast accuracy (see Figure 5.4). Usually, NWP model data and online measurement data are combined for the short-term forecast, giving more weight to the NWP data for longer forecast horizons and more weight to online data for shorter horizons. Very short-term forecasts of up to one or two hours are possible even without NWP model data. For forecast horizons of more than approximately half a day, the online data usually do not add information to the NWP model data and the short-term forecast ends.

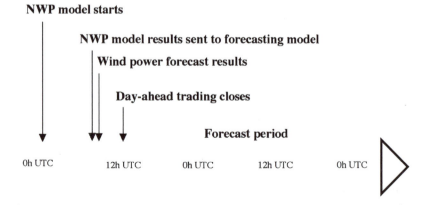

Figure 5.3 *Typical time schedule for wind power forecasting used for day-ahead trading*

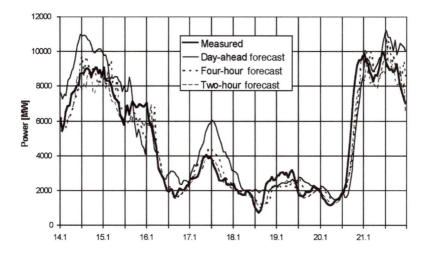

Figure 5.4 *Example time series of online measurement and forecasts of wind power generation in Germany; forecasts with different forecast horizons are shown*

REGIONAL UPSCALING

A wind power forecast for a larger region with many wind farms is usually made by forecasting only some of the wind farms and by extrapolating their power output to the whole region – often called regional upscaling. This minimizes the effort involved in making the forecasts and reduces the amount of data needed from NWP models as input. The accuracy of the forecasts does not decrease much since wind farms close to each other show a similar behaviour. However, it is important that the wind farms selected for forecasting are representative of all wind farms to which their output is extrapolated.

Different algorithms can be used for upscaling. Their main function is to calculate the output of all wind farms of the area from the known (or forecast) output of the representative wind farms. In the Wind Power Management System (WPMS), developed by ISET in Germany, the following mechanism is used. The area of interest is subdivided into grid squares. For each of these grid squares, the installed capacity of wind farms, their coordinates and hub heights, and the roughness of the terrain are known. This information is compiled from a database of all wind turbines in Germany, which includes:

- installed power;
- rotor radius;
- hub height;
- location;
- turbine type;
- surface roughness;
- date of erection and dismantling.

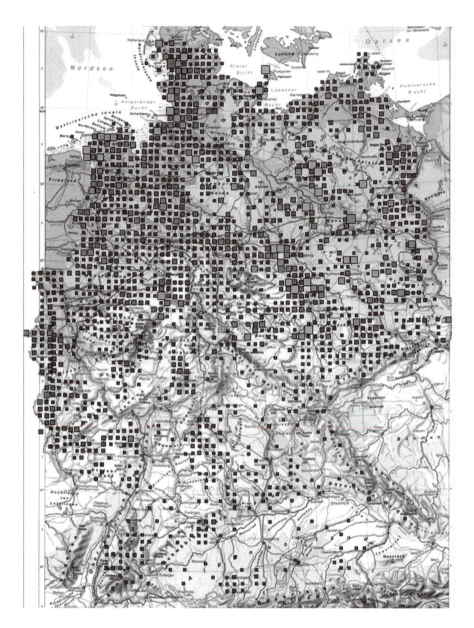

Figure 5.5 *Grid squares used by the Wind Power Management System for regional upscaling*

Figure 5.5 shows the grid squares for Germany as an example. The size of the squares shows information about installed capacity: the smallest squares are 1 megawatt (MW) to 13MW, and the largest squares are 131MW to 146MW. The WPMS calculates the power output of the wind turbines in each grid square by using the forecast power output of the representative wind farms.

The closer a reference wind farm is to the grid square, the greater is its influence (see Figure 5.6). Considering a case with i grid squares and j representative wind farms, the power output of the whole region P_{total} is the sum of the power output P_i from all its grid squares:

$$P_{total} = \sum_i P_i. \tag{1}$$

The power output of one grid square is calculated from the weighted power outputs of all representative wind farms:

$$P_i = k_i \sum_j A_{ij} P_j. \tag{2}$$

Here, P_j is the power output of representative wind farm j, normalized with its installed power, and k_i is a normalization factor. The weighting factors A_{ij} are calculated as:

$$A_{ij} = \exp\left(\frac{-S_{ij}}{S_0}\right) P_{IP,i} \tag{3}$$

where $P_{IP,i}$ is the installed power in grid square i, S_{ij} is the distance between a representative wind farm and the grid square, and S_0 is a spatial correlation parameter that has to be determined empirically.

The normalization factor k_i makes sure that the sum of all weighting factors equals 1:

$$k_i = \frac{1}{\sum_j A_{ij}}. \tag{4}$$

SMOOTHING EFFECT

The power output of wind farms fluctuates. These fluctuations are very difficult to forecast, and even if the power output on a particular day is predicted well, the fluctuations will cause a forecast error. The larger the wind farm, the smaller will be the fluctuations and the corresponding forecast error. If many wind farms are forecast together, the forecast error decreases further. In addition, the aggregation of large regions with several gigawatts of installed capacity will lead to a decrease in the relative forecast error since there will be cases where the forecast errors of different regions will partly cancel each other out. An example of this is shown in Figure 5.7, which shows the forecast error for the three German control zones with large wind power capacity: those of E.ON, VE-T and RWE, together with the error of the aggregated forecast for an example time series of four days. It can be seen that the forecast error for the aggregated wind power always stays below 2.5 per cent, while the error for

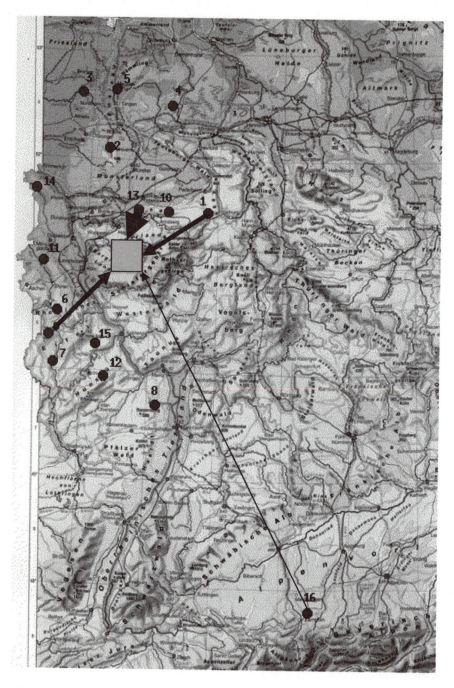

Figure 5.6 *Sketch of the calculation of the power output of a grid square and the upscaling mechanism*

single control zones reaches up to 8 per cent. The forecast error is given here as the difference between forecast and measurement as a percentage of the installed capacity.

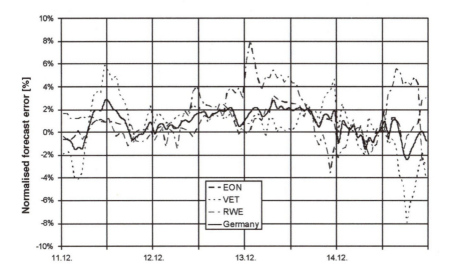

Figure 5.7 *Example time series of relative forecast error for the individual control zones of E.ON, VE-T and RWE, and for the whole of Germany*

The forecast error depends upon the number of wind turbines and wind farms and their geographical spread. In Germany, typical forecast errors for the representative wind farms forecast with WPMS are 10 to 15 per cent root mean square error (RMSE) of installed power, while the error for the control zones calculated from these representative wind farms is typically 6 to 7 per cent, and that for the whole of Germany only 5 to 6 per cent.

FORECAST ACCURACY

The accuracy of a wind power forecast is, of course, the most important criterion for its quality and value. Figure 5.8 shows an example time series of the day-ahead forecast for Germany, together with its monitored values for one month.

Since the forecast accuracy changes with time, a long time period has to be considered in order to evaluate the quality of a forecasting system. Since this is difficult with a time series plot, a scatter plot is often used. However, the information at which time a certain error occurred is lost in this evaluation method. An example of a scatter plot of forecast errors is given in Figure 5.9. The forecast wind power output for Germany is shown versus the monitored values. The data comprise a period of one year and are normalized with respect

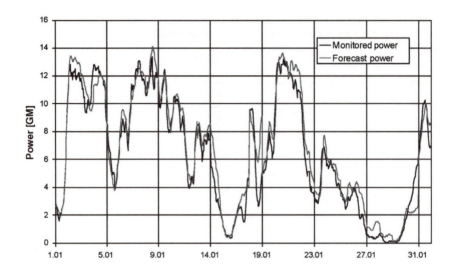

Figure 5.8 *Example time series of monitored and forecast power output for Germany*

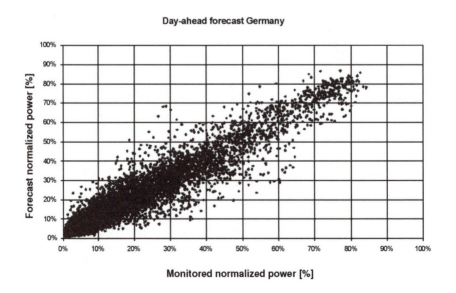

Figure 5.9 *Forecast versus monitored wind power output for Germany; values are normalized with respect to the installed capacity*

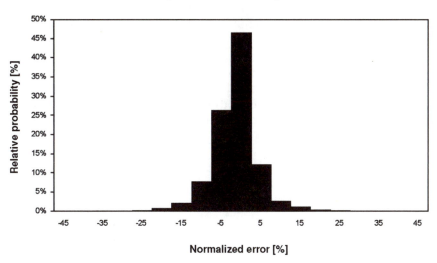

Note: Data are equivalent to data in Figure 5.9.

Figure 5.10 *Frequency distribution of the difference between forecast and monitored power output*

to the installed capacity. The forecast data are from a day-ahead forecast performed with ISET's Wind Power Management System using NWP data from the German Weather Service (Deutscher Wetterdienst, or DWD).

The information given in the scatter plot can be further condensed by calculating a frequency distribution of the forecast error. Here the information concerning at which power output a certain error occurred is not visible any more. Figure 5.10 shows an example using the same data as in the scatter plot.

Frequently, the information about the forecast error needs to be further condensed to only one or a few values. Many different error measures can be used for this:

- mean error (bias);
- mean average error (MAE);
- root mean square error (RMSE);
- correlation coefficient (*r*).

Additionally, different ways of relating the error to the production or size of the installation are used:

- normalized with respect to installed power;
- normalized with respect to mean power generation;
- normalized with respect to current power generation.

It has to be stressed that different measures lead to very different values. For comparison of different wind power forecasting systems, it is therefore extremely important to use the same error measures. Furthermore, the error depends upon many other influences, which have to be equal for a comparison of different systems:

- The error is different for each wind farm, depending upon local conditions, the size and location of the wind farm, geographical spread, etc.
- For regional forecasts, the error depends upon the number of wind farms, and their size and spatial distribution (see the section on 'Smoothing effect').
- The error depends upon the weather prediction model used as input.
- The error is different for different time periods.
- The error depends upon the amount and quality of the measured data used as input to the system.
- Finally, the error also depends upon the forecast horizon (see the section on 'Forecast horizon').

EXAMPLE: THE WIND POWER MANAGEMENT SYSTEM (WPMS)

Wind power forecasting is an integral part of the electricity supply system in Germany. The Wind Power Management System developed by ISET is used operationally by three of the four German transmission system operators (see Figure 5.11). The system consists of three parts:

1 the online monitoring, which performs an upscaling of online power production measurements at representative wind farms to the total wind power production in a grid area;
2 the day-ahead forecast of the wind power production by means of artificial neural networks (ANNs). This is based on input from a numerical weather prediction (NWP) model;
3 the short-term forecast, which also employs online wind power measurements to produce an improved forecast for up to eight hours ahead.

For a short-term wind power forecast, representative wind farms or wind farm groups have to be determined and equipped with online measurement technology. For the day-ahead forecast, only an historical time series of measured power output of the representative wind farms is needed. For these locations, forecast meteorological data obtained from a numerical weather prediction model are used as input. The resolution of the forecast and the forecast horizon depends upon the NWP data used. In Germany, an hourly resolution and a forecast horizon of three days are currently in operation.

Artificial neural networks are used to forecast the wind power generated by a wind farm from the predicted meteorological data of the NWP model. The ANNs are trained with NWP data and simultaneously measured wind

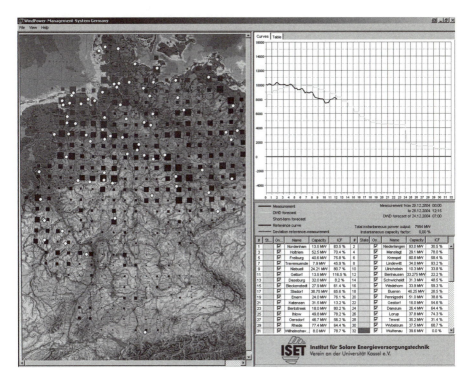

Figure 5.11 *The graphical user interface of the Wind Power Management System*

farm power data from the past in order to 'learn' the dependence of the power output upon predicted wind speed and additional meteorological parameters (see Figure 5.12). The advantage of an artificial neural network over other calculation procedures is that it 'learns' connections and 'conjectures' results, even in the case of incomplete or contradictory input data. Furthermore, the ANN can easily use additional meteorological data such as air pressure or temperature to improve the accuracy of the forecasts. In the operational forecast system, the deviation (RMSE as a percentage of the installed capacity) between the (day-ahead) predicted and actual occurring power for the control areas of E.ON, VE-T and RWE is currently about 6 to 7 per cent of the installed capacity. The forecast error for the total German grid amounts to 5 to 6 per cent.

In addition to the forecast of the total output of the wind turbines for the following days (up to 72 hours), short-term (15 minutes to 8 hours) forecasts are the basis for efficient and safe power system management. Apart from the meteorological values, such as wind speed, air pressure and temperature, online power measurements of representative sites are an important input for the short-term forecasts. As in the day-ahead forecast, ANNs are used to relate the input values to the power output. The forecast uncertainty is considerably lower than for the day-ahead forecast. For the German grid, the RMSE as a

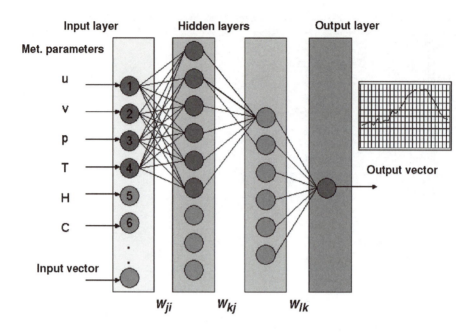

Figure 5.12 *Sketch of an artificial neural network (ANN) used for the wind power forecast*

percentage of the installed capacity is currently 2.6 per cent for the 2-hour-ahead forecast, and 3.6 per cent for the 4-hour-ahead forecast (ISET, 2005).

'LEARNING CURVE' OF FORECASTING ACCURACY

Since the WPMS forecasting system was first implemented in 2001, it has been constantly improved. The result is a continuous reduction of the forecast error, resulting in a 'learning curve' of decreasing forecast error over time, as can be seen in Figure 5.13, which shows the development of the forecasting error for the example of the E.ON control zone (Lange et al, 2006). The accuracy of the operational wind power forecast has improved from approximately 10 per cent RMSE at the first implementation in 2001 to an RMSE of about 6.5 per cent in 2005.

The operational experience of several years shows that the system has not only performed well in terms of accuracy, but also in terms of practical usability. The system has been installed in three different control-room software environments. It has been extended constantly to include user requirements and wishes, and now includes, for example, a hot standby capability with full monitoring and different options for the graphical user interface.

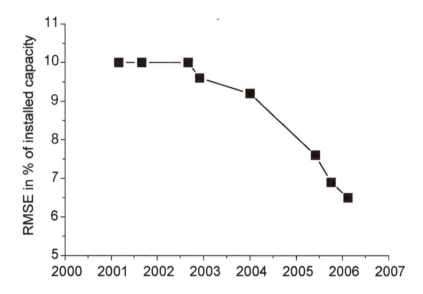

Figure 5.13 *Development of the forecasting error of the operational day-ahead forecast for a control zone; the root mean square error of the forecast time series is compared to that of the online monitoring*

EXAMPLES OF CURRENT RESEARCH

Improved representation of the atmospheric boundary layer

The selection of the input parameters for the ANN is of crucial importance for the performance of the forecast. Wind velocity and wind direction are, of course, the most important parameters for the wind power forecast. However, with the neural network approach, it is easily possible to incorporate additional parameters. The set of meteorological parameters used for the forecast has been improved to take into account the influence of atmospheric stability, especially for new turbines with high towers. This has led to an important improvement in forecast accuracy. Most important was the inclusion of the wind speed predicted by the NWP at 100m in height (ISET, 2005). As can be seen in Figure 5.14 for the example of one German Transmission System Operator (TSO) control zone, the forecast error (RMSE as a percentage of installed capacity) was reduced by more than 20 per cent. Two different numerical weather prediction models were used as input for the forecast, showing very similar results.

Multi-model approach for forecasting methods

The day-ahead wind power forecast by ANN using one method of artificial intelligence is used operationally by German TSOs. To improve the forecast

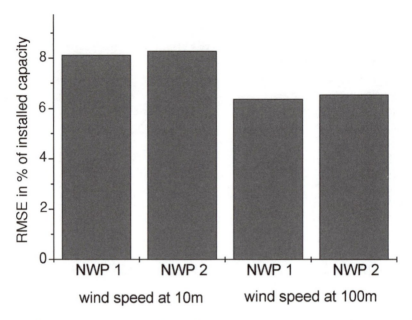

Figure 5.14 *Comparison of the wind power forecast accuracy for a control zone using 10m and 100m in height wind speed as the input parameter*

ability, other types of AI models were investigated in a comparative study (Jursa et al, 2006). In detail, these were:

- artificial neural networks (ANNs) as a reference;
- a mixture of experts (ME);
- nearest-neighbour search (NNS) combined with particle swarm optimization (PSO);
- support-vector machines (SVMs);
- a built ensemble comprising all models.

The ANN consists of nonlinear functions *g*, which are combined by a series of weighted linear filters (Gershenfeld, 1999). Here, a neural network with one 'hidden layer' with *j* 'neurons' was used, constituting the weight matrices *A* and *a*:

$$\hat{P}_i = g\left[\sum_{j=1}^{m} a_j \, g\left(\sum_{k=1}^{m} A_{jk} \, W_{kt}\right)\right].$$ [5]

The vector W_{kt} contains the input data from the numerical weather prediction model (i.e. *k* values of meteorological parameters at time *t*). P_i denotes the output value (i.e. the predicted power output of a wind farm at the time *t*).

The ME model is a construction of different 'expert' neural networks in order to tackle different regions of the data, using an extra 'gating' network, which also sees the input values and weights the different experts corresponding to the input values (Bishop, 1995).

The nearest-neighbour search (Hastie et al, 2001) uses those observations in an historical NWP data set closest in input space to the actual input values in order to form the output. The NNS method used is based upon the construction of a common time delay vector of weather data from several prediction locations of the NWP and upon an iterative algorithm consisting of the NNS and a superior PSO for the selection of optimal input weather data (Jursa et al, 2006).

The support-vector machine maps the input data vectors, W_t, into a high-dimensional feature space by calculating convolutions of inner products using support vectors, W_i, of the input space:

$$f(w_i) = sign \left[\sum_{support\ vectors} P_i\ \alpha_i\ K(w_i,\ w_t) - b \right]. \qquad [6]$$

In general, support-vector machines are learning machines that use a convolution of an inner product, K, allowing the construction of non-linear decision functions in the input space, which are equivalent to linear decision functions in the feature space. In this feature space, an optimally separating hyper-plane is constructed (Vapnik, 2000).

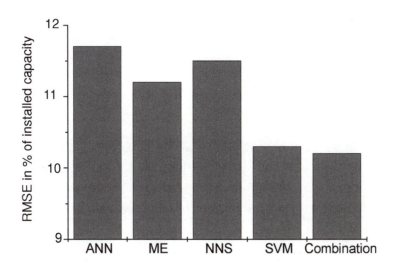

Note: The methods used are artificial neural networks (ANNs), a mixture of experts (ME), nearest-neighbour search (NNS) and support-vector machine (SVM).

Figure 5.15 *Comparison of the mean root mean square error of a wind power forecast for a group of single wind farms obtained with different artificial intelligence methods and with a combination of all methods*

A comparative study between the different forecasting methods has been performed using the power output measurements of ten wind farms in the E.ON control zone and corresponding NWP prediction data for these points from the German Weather Service. Data from September 2000 to July 2003 have been used. Figure 5.15 shows the comparison of the mean RMSE for the ten wind farms. It can be seen that, in this case, the support-vector machine yields the best results. In addition, a simple ensemble approach has been tested by averaging the outputs of the models studied. As can be seen in Figure 5.15, even this simple ensemble improves the forecast accuracy compared to the results of the single ensemble members.

Multi-model approach for numerical weather forecast models

A study has been performed to investigate the influence of merging different NWP models on the accuracy of the wind power forecast. Three different NWP models have been used for a day-ahead wind power forecast for Germany (see Table 5.2). All three models have been used as input to the WPMS based on the ANN method. Network training was performed with data of more than one year. A concurrent data set of seven months (April to October 2004) has been used for the comparison.

The RMSE percentage of the installed capacity of the three models is shown in Figure 5.16. It can be seen that the differences between the models are minimal. Additionally, a simple combination of the three models has been tested by averaging their forecasts. It is clear that even this simple approach improves the forecast accuracy very significantly compared to the results of the single models. The resulting RMSE for the combined model for Germany is 4.7 per cent, while the values for the individual forecasts are between 5.8 and 6.1 per cent.

Table 5.2 *Main characteristics of the numerical weather prediction (NWP) models used*

	NWP 1	NWP 2	NWP 3
Forecast schedule	72 hours	48 hours	72 hours
Model runs	00 and 12 coordinated universal time (UTC)	00 UTC	00 UTC
Available parameters	Wind speed Wind direction Temperature Air pressure Humidity	Wind speed Wind direction Temperature Air pressure Humidity Momentum flux	Wind speed Wind direction Temperature Air pressure

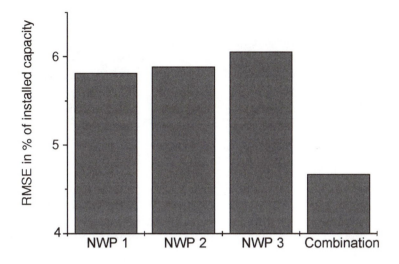

Figure 5.16 *Comparison of the root mean square error of a wind power forecast for Germany obtained with the Wind Power Management System based on artificial neural networks, with input data from three different numerical weather prediction models and with a combination of these models*

Prediction of the forecast uncertainty

In addition to the wind power forecast itself, it is important to know the uncertainties of this forecast. The forecast's confidence interval gives a quantitative measure of the possible deviation of the actual wind power from the forecast depending upon the meteorological input data for each time step. A statistical method has been used to predict not only the power output, but also an upper and lower limit for the forecast accuracy for each time step (see Figure 5.17) (ISET, 2005). The method is based on determining the forecast uncertainty for each representative wind farm, depending upon wind speed and wind direction. The total uncertainty is then calculated from the uncertainty estimations of all representative wind farms.

FUTURE CHALLENGES

As wind power capacity grows rapidly in Germany and in many other countries, forecast accuracy becomes increasingly important. Especially in relation to large offshore wind farms, an accurate forecast is crucial due to the concentration of large capacity in a small area. However, during recent years, forecast accuracy has improved constantly, and it can be expected that this improve-

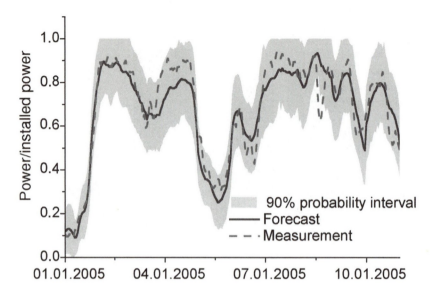

Figure 5.17 *Example time series of the forecast power output and its 90 per cent probability interval, compared with the values of the online monitoring*

ment will be maintained in the future. For the Wind Power Management System, a number of improvements are planned:

- The development of operational ensemble model systems using the data from several numerical weather prediction models will clearly improve forecast accuracy. In addition, an improved method for model combination will be developed.
- Improvements in the numerical weather prediction models and more frequent updates of weather predictions will enhance the input data for wind power forecasting.
- Further improvements in the forecasting methods and improved methods for the combination of different forecasting methods can be expected to further reduce forecasting errors.
- Especially for short-term wind power forecasting, additional use of online wind measurement data has the potential to improve forecasts. In Germany, the ISET wind measurement network (Hahn et al, 2006) will be used to correct the NWP data used for forecasts.

Forecast accuracy is only one of the challenges for future wind power forecasting systems. The scope of systems will also have to be extended in order to meet future challenges:

- Wind power forecast in the offshore environment has the potential to become more reliable than on land if specific offshore forecast models are developed. The meteorological situation in the near-shore marine atmospheric boundary layer differs from that over land. Atmospheric stability and distance to the shore are particularly important.

- Improved forecasts for short time horizons will be needed for grid safety and intra-day trading.
- Predicting the probability distribution of the forecasting error and minimizing events with large errors provide opportunities of reducing the reserve capacity for balancing wind power forecast errors.
- Forecasts in high spatial resolution for each grid node of the high voltage grid will be needed for high wind power penetration in order to tackle the problem of congestion management.

ACKNOWLEDGEMENT

This work on this chapter has been partially funded by the European Commission in the DESIRE (Dissemination Strategy on Electricity Balancing for Large-Scale Integration of Renewable Energy) project (TREN/05/FP6EN/S07.43516/513473). The research described has been partially funded by the Federal Ministry for the Environment, Nature Conservation and Nuclear Safety (Project no 0329915A).

REFERENCES

Bishop, C. M. (1995) *Neural Networks for Pattern Recognition*, Oxford University Press, Oxford

Doms, G. and Schättler, U. (1999) *The Non-Hydrostatic Limited Area Model LM (Lokal-Modell) of DWD. Part I: Scientific Documentation*, Deutscher Wetterdienst, Geschäftsbereich Forschung und Entwicklung, Offenbach, Germany, www.cosmo-model.org

FLaP (2002) *FLaP Computer Program*, University of Oldenburg, Oldenburg, Germany

Focken, U., Lange, M. and Waldl, H. P. (2001) 'Previento: A wind power prediction system with innovative upscaling algorithm', in *Proceedings of the European Wind Energy Conference*, Copenhagen, p826ff

Focken, U., Lange, M. and Waldl, H. P. (2002) 'Previento: Regional wind power prediction with risk control', *Proceedings of the Global Wind Power Conference*, Paris

Gershenfeld, N. (1999) *The Nature of Mathematical Modelling*, Cambridge University Press, Cambridge

Giebel, G., Brownsword, R. and Kariniotakis, G. (2003) *The State-of-the-Art in Short-Term Prediction of Wind Power: A Literature Overview*, Report of the Anemos project, www.anemos.cms.fr

Grell, G. A., Dudhia, J. and Stauffer, D. R. (1994) *A Description of the Fifth-Generation Penn State/NCAR Mesoscale Model (MM5)*, Technical Report NCAR/TN-398+STR, National Center for Atmospheric Research, Boulder, Colorado

Hahn, B., Durstewitz, M. and Döpfer, R. (2006) 'ISET Wind Energy Measurement Network', in *Proceedings of the European Wind Energy Conference 2006*, Athens, Greece

Hastie, T., Tibshirani, R. and Friedman, J. (2001) *The Elements of Statistical Learning: Data Mining Inference and Prediction*, Springer Verlag, Berlin

ISET (2005) *Entwicklung eines Rechenmodells zur Windleistungsprognose für das Gebiet des Deutschen Verbundnetzes*, Report, ISET e.V., Kassel, Germany

Jursa, R., Lange, B. and Rohrig, K. (2006) 'Advanced wind power prediction with artificial intelligence methods', in *Proceedings of the First International ICSC Symposium on Artificial Intelligence in Energy Systems and Power*, 7–10 February Madeira, Portugal

Landberg, L. (1994) *Short-Term Prediction of Local Wind Conditions*, Technical report, Risø-R-702(EN), Risø National Laboratory, Denmark

Landberg, L. (2001) 'Short-term prediction of local wind conditions', *Journal of Wind Engineering and Industrial Aerodynamics*, vol 89, pp235–245

Landberg, L., Giebel, G., Nielsen, H. A., Nielsen, T. S. and Madsen, H. (2003) 'Short-term prediction: An overview', *Wind Energy*, vol 6, no 3, June, pp273–280

Lange, M. and Focken, U. (2005) *Prediction of Wind Power and Reducing the Uncertainty for Grid Operators*, Second Workshop of International Feed-In Cooperation, 15–16 December, Berlin

Lange, B., Rohrig, K., Ernst, B., Schlögl, F., Cali, Ü., Jursa, R. and Moradi, J. (2006) 'Wind power forecasting in Germany: Recent advances and future challenges', *Zeitschrift für Energiewirtschaft*, vol 30, no 2, pp115–120

Mortensen, N. G., Landberg, L., Troen, I. and Petersen, E. L. (1993) *Wind Atlas Analysis and Application Program (WAsP)*, Risø National Laboratory, Roskilde, Denmark

Nielsen, T. S. and Madsen, H. (1997) 'Statistical methods for predicting wind power', in *Proceedings of the European Wind Energy Conference*, October, Dublin, pp755–758

Nielsen, T. S., Nielsen, H. A. and Madsen, H. (2002) 'Prediction of wind power using time-varying coefficient-functions', in *Proceedings of the XV IFAC World Congress*, Barcelona

PARK (1993) *PARK Computer Program*, Risø National Laboratory, Roskilde, Denmark

Rohrig, K. and Ernst, B. (2000) 'Online-supervision and prediction of 2500 MW wind power', in *Proceedings of the EWEA Special Topic Conference: Wind Power for the 21st Century*, Kassel, Germany

Vapnik, V. N. (2000) *The Nature of Statistical Learning Theory*, 2nd edn, Springer Verlag, Berlin

6

Flexibility of Fossil Fuel Plant in a Renewable Energy Scenario: Possible Implications for the UK

Fred Starr

INTRODUCTION

As other authors in this book have made clear, it is beginning to be accepted that by 2020 as much as 20 per cent of the electricity in the UK could come from renewable sources, much of this derived from the generation of power from wind (Environmental Audit Committee, 2006). It is also recognized that in some manner, the non-renewables sector (i.e. fossil fuel and nuclear plants) will have to compensate for the irregularity or intermittency that is a feature of this form of power generation. The aim of this chapter is to review how intermittency could affect the design and operation of future power plants, whether they utilize coal, gas or nuclear energy for electricity generation.

In discussing how intermittency will affect power plants, a number of issues have to be considered. What is the time scale under review, bearing in mind that planning, design and construction can take up to a decade before a new plant begins to generate power? Is the assessment to be based upon the situation as it is at present, when the main question is how well existing coal- and gas-fired power plant can cope? Or is it set 15 to 20 years in the future, when new designs of fossil and nuclear plant will be coming into operation? These, one presumes, will have been designed with global warming in mind. Will these more advanced concepts be better or worse than current types of UK plants at coping with the problem of intermittency?

The intermittency of wind generation brings a number of problems with which the rest of the system will have to contend. The most obvious need is for large amounts of power to become available to the grid when the wind has dropped for a long period. This requires plants that can be brought up to full power within a matter of hours, without damaging them in the process. The converse situation is for generating plant to be capable of being shut down quickly, without too much wastage of fuel, when power from wind has been restored. Even when the wind is blowing, there will be short-term fluctuations.

These must be compensated for by plants that can change their power output rapidly. It does not necessarily follow that a generating station that can be quickly started up or shut down will necessarily have the ability to respond to fluctuations in the load. Related to this problem, short-term variations in supply and demand will lead to irregularities in voltage and frequency. As will be described, large-scale gas turbine-based, combined-cycle plants are not necessarily the best at compensating for this sort of problem.

In evaluating the importance of the intermittency issue, it must be kept in mind that the number of centralized plants in the UK is tending to fall simply because the output of modern plants is very large. A unit of 500MW is not exceptional. One such unit will supply the peak demands of over 150,000 households, or a town about the size of Nottingham. A modern power station site usually contains a number of these very big units, all of the same design, which in terms of being able to respond to intermittency will behave in the same way. They will all be feeding into the same point of the grid, and any shortcomings in the way that they operate will have a major effect. The future choice of generating plant therefore requires even more consideration than it does at the present time.

One argument that has been advanced to downplay the impact of wind-generated power is that the UK has a 'pseudo-intermittency' problem already due to the day-to-night variation in electrical demand. It should be noted that this diurnal variation has become less significant over the decades. The day-to-night difference is now less than a factor of 2; during the 1960s, it was 3, and it was close to 5 during the 1930s – so this type of intermittency tends to be of reduced importance. Furthermore, improved predictive techniques have enabled the UK National Grid to make an extremely good assessment of power demands; as a result, power station operators know how likely it is that they will need to generate electricity.

For fossil fuel plant operators, working against this is the fact that much of the base-load power in the UK now comes from nuclear sources. The additional daytime load is taken up by coal-fired steam plant and by combined-cycle gas turbines (CCGTs) fuelled by natural gas. The 'peakiness' of fossil plant generation increases maintenance costs, requires extra fuel and can result in what is known as 'forced outages', where a plant has to shut down without warning because of the failure of some vital piece of equipment. Depending upon the regulatory regime, forced outages can cost an operator prohibitive amounts of money in having to buy backup power from its competitors.

Looking to the future, although the day-to-night variation in *demand* from consumers is not likely to change much, the smooth variation in *output* from the conventional generating sector is likely to disappear as more and more wind and solar energy is brought onto the system. Indeed, judging from German predictions, admittedly set around 2050, there could be periods when there is zero demand from fossil plant for days at a time (Quaschning, 2001).

As a result, a large increase in renewables will start to increase the peak-to-trough ratio in power station use, and this needs to be taken into account when the future mix of power plants and the specific designs are being considered. Accordingly, this chapter will briefly review the situation as it is now, then look at what this might imply in terms of plant designs that could be more suitable for a scenario in which much more power comes from intermittent sources.

THE UK POWER PLANT SYSTEM OF TODAY

Whenever the subject of intermittency is considered, thinking is dominated by the UK power system as it is currently configured, and in which renewables play little part. At the present time, coal-fired steam plant and natural gas-fuelled combined cycle together supply about 65 to 70 per cent, and nuclear about 20 to 25 per cent of the power in the UK. But the UK energy sector has been evolving and continues to evolve. The power industry has responded to changes to increased demand for electricity, but also in the pattern of demand. For example, the reduction of the day-to-night ratio in electrical consumption favours the construction of plants that are optimized for base-load operation.

The evolution of the power system has also been strongly dependent upon the availability of fuels. In recent decades, the most noticeable change has been the move away from coal to natural gas, and the rise of the CCGTs. Apart from the nuclear plant Sizewell B, the CCGTs have been the only new construction in the UK since 1986, when the last unit of the Drax coal-fired steam complex was brought into commission.

Economic considerations imply that, at any one time, the power industry needs to make the best use of existing equipment. Although obsolescent in terms of efficiency, the nation's coal-fired units are able to compete because of a number of factors. Plant costs have been written down, coal prices are low compared to natural gas, and manning levels have been reduced. Helping to preserve the continued operation of coal-fired steam plant is the perception that these are better than CCGTs in giving stability to the grid. But there must come a time when, because of wear and tear, it will be impossible to run these coal-fired units profitably. Any new plants, especially those burning coal, will need to be more efficient and should be of the 'clean fossil' type, which will help to reduce fuel use and cut down on carbon dioxide (CO_2) emissions.

The situation with respect to nuclear energy in the UK is not dissimilar, with most of the current plants reaching the end of their life (DTI, 2006a). There are three Magnox stations still operating; but all are scheduled to be closed down by 2012. The existing Advanced Gas-cooled Reactor (AGR) plants will have to be decommissioned by the early 2020s. Although more efficient than standard Pressurized Water Reactors (PWRs), the graphite cores and high temperature superheaters that are a feature of AGRs have a definite life. Because of their high temperature operation, the Magnox and AGRs have to be run as base-load power units. Only Sizewell B, a PWR, is likely to be left in service and could be operating until the middle of the century. Although it is not really clear what sort of nuclear plants will be built in the future, the emphasis seems to be on improved forms of PWR, which for both economic and technical reasons will be intended for base-load operation.

This is the broad picture of the present and medium-term future. However, any new centralized plants will need to work alongside other sources of electricity, besides renewables. Partly in the interests of energy saving, and partly to help reduce CO_2 emissions, the UK Government supports the European Union (EU) Cogeneration Directive (European Parliament, 2004), which should lead to a significant amount of electricity being generated from Combined Heat and Power (CHP) plants. Whether CHP will help with the intermittency issue is debatable (DTI, 2006b). Much will depend upon whether cogeneration sets can produce extra electricity when power is short, or whether they are able to operate as heat-only plants when there is a surplus of power. Experience from other countries may not be too helpful since, in the author's view, the biggest growth in cogeneration in the UK is likely to come from micro-CHP. In contrast to district heating-style CHP, micro-cogeneration 'sets' will be integrated within household gas boilers, producing between 1kW and 3kW (Pehnt et al, 2005). If these are to give support to the grid, cheap and efficient electronic systems will be needed if good quality power is to be safely exported to the grid from millions of units. There is a real need for development in this area.

So, what are the implications for the design and construction of advanced centralized power plant in which wind and eventually solar energy will be taking over much of the demand? Although some new nuclear capacity will probably be built, the focus of this chapter is on coal- and gas-fuelled power plants. Only these are capable of meeting rapid variations in demand. They have this ability at present, and are used to stabilize grid frequency and voltage; but will the newer fossil fuel plants be able to cope as well? And if not, why should this be and what might be done about it?

ADVANCED GENERATING PLANTS, ENERGY SAVINGS AND THE ISSUE OF CLIMATE CHANGE

New coal- and gas-fired plants should be more efficient and cheaper to run than the ones they replace. In addition, they will have to meet some new requirements. Their greenhouse gas emissions, principally CO_2, will have to be significantly lower than the plants of today. This is easy for nuclear, but more difficult for fossil-fuelled generating systems. The necessary cuts in CO_2 emission from fossil plants cannot simply be met by improvements in efficiency. In the UK, the construction of a new set of coal-fired stations would cut emissions by 50 per cent over current levels. This sounds very good; but, as noted earlier, our present 'fleet' of coal-fired steam plants are obsolescent by world standards. In contrast, since most of the natural gas-fuelled CCGTs that are operating in the UK are relatively new, the CO_2 savings that could be made by replacing the present set are probably not much more than 10 per cent. If CO_2 emissions from fossil fuel plants are to be brought down to acceptable levels, any new large generating plants, helping to give stability to the grid, will also have to 'capture' the CO_2 for storage in disused oil or gas fields.

The need for such units to be of the carbon capture type will have significant implications for their design. The process of capturing CO_2 will absorb much power. It is therefore vital that the thermodynamic efficiency of such plants should be as high as possible. High efficiency reduces CO_2 emissions in terms of tonnes of CO_2 per megawatt hour (MWh), and reduces the amount of CO_2 that needs to be captured. In other words, high efficiency increases the power output from a plant so that the impact of parasitic losses associated with capturing CO_2 is reduced.

DESIGN AND OPERATION OF COAL- AND NATURAL GAS-POWERED STEAM PLANTS

To understand how the issue of intermittency will affect power plants, it is necessary to have a reasonably good idea of how modern coal-fired steam plant and natural gas-fired CCGT plants are designed and operated. The main points are covered below; but the book *Steam: Its Production and Use*, which has been updated over the years, is very comprehensive (Babcock and Wilcox, 2004). The UK Department of Trade and Industry (DTI) has also published a number of summaries on modern power plants that are extremely helpful (DTI, 1999, 2004, 2006c).

Steam plants

The oldest type of power plant is that of the fossil fuel steam plant, in which coal or some other fossil fuel is burned in a boiler, generating steam to drive steam turbines. The furnace boiler is very roughly the size and shape of a high-rise apartment block, about 40m in height. In the UK, the boiler, along with the steam turbines and alternator, are housed in very large buildings, the height of which is governed by the size of the boiler. The new Tate Modern Gallery in London, for example, was formerly Bankside Power Station.

The interior walls of the boiler are lined with tubes in which steam is generated. The temperatures within the furnace itself are in the 1500° C to 2000° C range, and the combustion products or 'flue gases' at the start of the furnace exit duct are at around 1200° C. There is still a great deal of heat left in the flue gases, which after entering the exit duct are used to superheat and reheat the steam, to preheat the boiler water, and to preheat the air required for combustion of the fuel. The water and steam in the boiler are at 150 to 300 bar pressure, depending upon how advanced the design is. The pressure vessels, connecting pipe work and tubing have to be very thick walled to contain the water and steam, even though high strength steels are used in their construction.

The steam that is generated from the boiler cannot be used directly. It emerges from the boiler at about 300° C to 350° C, and needs to be heated to around 550° C before it can be used in the steam turbines. The temperature increase occurs in the superheater, which consists of arrays of tubes that are set across the flue gas duct. After superheating, the steam is piped across to the high pressure turbine, where it causes the turbine to rotate. Expansion of the steam through the turbine results in the pressure and temperature dropping to about 40 bar and 350° C. If the steam is to be used efficiently, it must be 'reheated' again to 550° C before being passed to the medium-pressure and, finally, the low-pressure turbines. At the exit of the low-pressure turbine, steam pressure and temperature are extremely low – circa 0.05 bar and about 35° C. At this point, the steam is condensed back to water. The condensed water is then pumped back to the boiler and the cycle starts again. The turbines, high pressure, and medium and low pressure are arranged in tandem, driving onto the same shaft to which the alternator is connected.

Changes in electricity output are primarily controlled by alteration of the water and fuel flows; but a change to these inputs will take time to work through, because of the amount of heat stored in the boiler, furnace and duct-work. Faster changes in turbine output are obtained by opening or closing the throttle at the inlet to the high-pressure turbine.

It is vital to ensure that the oxygen content of the boiler water is very low and the water extremely pure, otherwise the boilers and feed heaters will corrode. Control of water quality is not too difficult during normal operation; but during shutdown, air can leak into the steam system and, for various reasons, contamination of the boiler water becomes more likely. These problems can lead to the cracking of major pieces of equipment during repeated start-ups, because of a combination of bad water conditions and the thermally induced stresses.

During normal operation, the furnace tubing, furnace structure, super-heater, reheater and economizer run at high temperature. It will be apparent that it will take a long time to bring these up to temperature, the heat coming from the combustion of fuel. But there are many other pieces of equipment that run hot, such as valves for control of steam flow, connecting pipelines and feed heaters, all of which require steam to flow before these are at temperature. It will take several hours to get a plant to produce power from a completely cold start. Even if the plant has only been shut down overnight and is still quite warm when the restart commences, it will still take about an hour before electricity can be can be generated, and perhaps another one to two hours before the plant is up to full load. Shutdown is faster; but this too, must be controlled or the stored heat in the plant will be wasted. Excess steam during these periods will be 'dumped' in the condenser.

During these periods of temperature change, some parts of major components heat up faster than others, giving rise to differential expansion and high stresses. This problem was recognized by the Central Electricity Generating Board (CEGB), the original purchasers, who had the steam turbines of these plants designed for 5000 hot starts, 1000 warm starts and 200 cold starts. Nevertheless, premature failures can occur because of over-rapid start-ups and shutdowns, or because of poor detailed design. It is also well understood that as plant ages, the risks of these increase.

A low rate of start-up is a drawback of the steam plant of today; but more advanced plant may take even longer to restart, despite modifications to eliminate some of the present shortcomings. But once in full operation, steam plants are very efficient at controlling the grid frequency, voltage and load changes. This is particularly true of the older type of steam plant, such designs being of the 'steam drum' type, which has typified UK generating plant in the past.

The drum is a large pressure vessel about half full of very hot water, which is just on the point of boiling (note that the boiling point at steam plant pressures is close to 300° C!). Water from the drum is sent down to the inlets of the boiler tubes at the base of the furnace. On entering the tubes, the water picks up heat and begin to turns into steam; the mixture of steam and hot water rises up the tubes and returns back to the drum. In the steam drum, water boils off from the surface of the water and passes into the superheater. The significant advantage of this older type of plant is that if extra power is required, the drum acts as a reservoir, being full of hot water at boiling point. More steam can be produced by simply opening a valve. This ability is very useful if the grid starts to become overloaded, with the frequency and voltage beginning to drop. But the steam drum is a limiting factor in getting a plant online. The walls of the drum are some tens of centimetres in thickness, and frequent temperature changes will cause the drum to crack, given enough start-ups.

Drum-type designs in steam plant are now obsolete and have given over to 'once-through' boilers. Most modern European power plants are of this type. In once-through boilers, all the water turns into steam in the evaporator, passing straight into the superheater. As such, once-through designs can be brought online relatively quickly, although there can be a sudden temperature change early on in the start-up, at just about the point when the boiler begins to produce a significant amount of steam. There is also some evidence that the turn-down rate can be higher than with drum systems. Once-through operation becomes essential in what are termed supercritical and ultra-supercritical plants. In such units, there is no definite transition between water and steam, as conditions are above the supercritical pressure for water, which is 221 bar. In Europe, supercritical boilers are set to be the standard design because of the higher efficiencies: in the 43 to 46 per cent range for modern units.

Combined-cycle gas turbine generating plants

CCGTs consist of a gas turbine(s) that produces about two-thirds of the power from the plant. A single steam turbine set utilizes the steam generated in a Heat Recovery Steam Generator (HRSG) to produce the remaining third. The gas turbine operates by compressing air to a pressure of about 20 to 25 bar, burning natural gas in the compressed air and then expanding this through a combustion turbine, which then drives a compressor and an alternator. Some forms of CCGT have the set of steam turbines on the same shaft as the gas turbine. Only one alternator is then needed. This is more efficient, but less flexible in meeting changes in power demand than the other approach, which is to have an alternator for each gas turbine or steam turbine set.

The mixture of hot gases entering the combustion turbine is in the range 1250° C to 1450° C, depending upon the design; but by passing cooling air or steam through the turbines blades and other sections of the combustion system, metal temperatures are kept below about 950° C. Hence, the turbine components are subject to temperature gradients that change as the gas turbines start up and shut down.

These 'industrial' gas turbines are basically similar to the jet engines used on commercial aircraft. The main difference is that the power is used to drive an alternator, rather than being used to produce jet thrust via a ducted fan. A modern industrial turbine will, in energy terms, develop about four times as much power as even the most advanced jet engines and correspondingly much more massive. This brings bigger problems with component manufacture and greater susceptibility to temperature changes.

Turbine inlet temperatures are very critical. Even with advanced materials, a temperature increase of 20° C will halve the life of the turbine. There is usually a reserve when equipment first goes into service; but over the life of the gas turbine, compressor performance will deteriorate and this will lead to a steady increase in turbine temperatures, which can only be compensated for by cutting back on the fuel and reducing output. Over a much shorter period, compressors will become dirty and the deteriorating aerodynamics will also tend to increase turbine inlet temperatures. Fuel input and power output must be adjusted, and periodically the compressor must be washed to help restore its performance.

Some of the most advanced materials and components yet developed are used in the combustion section of the gas turbine. The turbine blades are one such example. Although of an aerodynamic shape, modern blades actually consist of a single crystal of a complex nickel-based alloy. The interior of the blades is interwoven with fine passages through which cooling air or steam passes. The cost of these blades is high; but in typical base-load operation, they will last at least 25,000 hours. Frequent stop–start operation will reduce this significantly.

The exhaust gas leaving the gas turbine is between 520° C and 640° C. Its heat is used for steam-raising in the HRSG. The HRSG supplies the steam for the steam turbines, which represent the other part of the combined cycle. As with coal-fired steam plant, water is evaporated to produce steam in the HRSG, which is then superheated. But there are big differences between the heating arrangements in an HRSG and those in steam plant. In the latter, furnace temperatures are very high, so there is no difficulty in raising steam or giving the steam the required amount of superheat. In an HRSG, at any point in the exhaust duct, the difference between the steam and gas turbine exhaust temperatures is quite small. To compensate for this, the HRSG has to be made disproportionately large. The cross-section of an HRSG duct would be about that of the frontage of a pair of semi-detached houses, and it could be 50m to 75m in length and height. Even so, steam temperatures tend to be closer to 500° C than 600° C, and pressures are currently below 165 bar (see Figure 6.1).

In the horizontal form of HRSG, the exhaust duct is laid out along the ground until it turns upwards to meet the stack. Six to eight 'harp-type' heat exchangers are placed across the duct to generate and superheat steam. In the vertical HRSG, the duct turns upwards soon after leaving the exhaust of the gas turbines (see Figure 6.2). Here again the heat exchangers are positioned across the duct; but the tubes are set in a series of loops. The advantage of the horizontal HRSG is that pumping power is less, and where ground space is available they can be cheaper to build. Vertical designs can have good condensate draining characteristics and have been promoted as being more suitable for stop–start operation. In practice, draining is not always as efficient as it might be. As will be described, the subject of draining is very important in HRSGs because of the strong likelihood of steam condensing in the system during normal plant shutdowns or plant trips (emergency shutdowns).

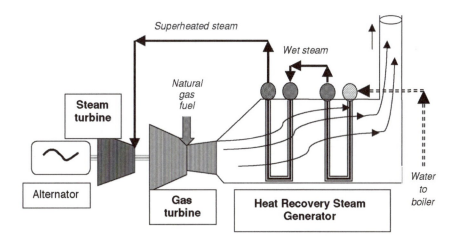

Source: Open University produced figure based on F. Starr sketch

Figure 6.1 *Schematic of a combined-cycle gas turbine with a horizontal form of heat recovery steam generator*

Each of these exchangers contains scores of tubes whose diameter is around 5cm, with each tube being positioned just a few centimetres from another. Apart from the first row of tubes (the one receiving a high thermal input from the gas turbine exhaust), all the tubes are set with closely pitched fins. Repair, if anything goes wrong, is extremely difficult.

The low temperature differences between gas turbine exhaust and steam require the steam to be produced at a minimum of two different pressures. The higher pressure is 100 to 160 bar, and the lower is 4 to 10 bar. Hence, an HRSG can be regarded as having two separate boiler systems. The heat exchangers associated with the high pressure boiler tend to be located in the hotter parts of the duct, close to the gas turbine exhaust. In both sets of boilers, water is brought up to temperature through an economizer before it is made to generate steam in the boiler or evaporator. The mixture of water and steam from the evaporators passes into the steam drum, where, as in the older type of steam plant, the steam is taken off the drum and passed into the superheater section. Once-through designs of HRSGs are still quite rare for the bigger plants, and it would appear that a standard approach has yet to be developed.

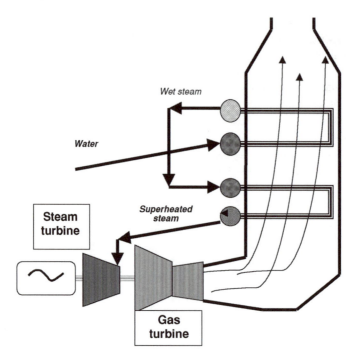

Figure 6.2 *Vertical heat recovery steam generator arrangement*

A big design issue is to ensure that each set of heat exchangers absorbs the correct amount of heat, otherwise some exchangers will run hotter than is desirable, whereas others will be starved of heat. This is more difficult than it might seem. The basic problem is a result of the fact that as water turns to steam, a great deal of heat has to be absorbed, but there is no increase in temperature. These heat transfer considerations have tended to dictate the design procedure, and this (as well as the need to reduce construction costs) has resulted in HRSGs that, in some cases, are not really suitable for stop–start operation.

PSEUDO-INTERMITTENCY WITH TODAY'S PLANTS

Load following and frequency control

As noted in the Introduction, we have one form of 'intermittency' that results from the day-to-night variation in the demand for electricity. In the UK, this demand is met in two ways. Nuclear plants, plus a small number of fossil fuel plants, provide the base load. Five years ago, combined-cycle gas turbine plants using natural gas provided the base load from the fossil fuel sector. Today, it is the most efficient coal-fired units that operate around the clock. On a typical day the combination of nuclear and coal, with some CCGT plants, might be supplying 20 gigawatts (GW) to 30GW of power.

Although there is a big difference between the day and night demand for power, which leads to some plants having to be shut down at night, the demand for power during the day is never constant. These particular variations in load can be met using coal-fired steam plants, the more efficient of which will operate on a 24-hour basis. Output from steam plants can be reduced to about 50 per cent of design without much difficulty. But when in load-following mode, plants are expected to maintain grid voltage and frequency at 230 volts and 50 hertz (Hz). As noted, this is not difficult with steam plant as the steam flow to the turbines can be quickly changed through actuation of the governors. In making these changes, temperature and pressure do not vary substantially. The main effect is that by operating at reduced output, efficiency goes down slightly. Pressurized water reactors should also be able to offer some help with frequency control; but if load following is needed from the nuclear sector, reactors of the boiling water type would seem to offer more capability. The UK has none of these.

The frequency issue is more problematic for CCGTs. A modern CCGT consists of a gas turbine and steam turbine, which are, in big modern units, connected by the same shaft, driving on to an alternator, which nominally runs at 3000 revolutions per minute. The alternator, and everything else in such a machine, is locked to the grid at the nominal frequency of 50Hz. If there is a sudden increase in demand on the grid, all the alternators throughout the grid network will slow down slightly until more power can be delivered. On a coal-fired steam plant, this is fairly easy; power can be increased by opening the throttle to the steam turbine. It is a bit more difficult for CCGTs. The slowing-down of the alternator will slow down the gas and steam turbines; unfortunately, the result is that the power output from the gas turbine drops. Less air is taken in by the gas turbine compressor and less fuel can be burned.

This is clearly a dangerous situation if one has a grid in which all the power is coming from CCGTs since the drop in grid frequency, caused by the demand for power, will lead to a drop, rather than an increase, in the power output. The situation is not quite as bad as it may seem since the voltage in the system will drop, and this will reduce the demand for power from many consumers. Many consumers will have experienced such 'brown-outs' when there has been a lack of capacity, perhaps due to the weather or industrial disputes.

Not all CCGTs are quite as susceptible as the big tandem shaft designs described above. Some 'merchant power' plants in the UK, designed for meeting varying loads, have independent alternators for the gas turbines and steam turbines. A typical set-up would be two gas turbines feeding into one HRSG, which supplies one steam turbine set. In such a case, the steam turbine and its alternator can be operated independently.

'Aero-derived' gas turbines, based on turbojets, intended for grid reinforcement and which are not of the CCGT type, do not suffer from the frequency problem, which is one reason why they are so useful for local reinforcement of the grid. They have completely separate power turbines to drive the alternator. Hence, if the grid runs slow or fast, there is no direct effect on the engine or its power output.

Whereas in a coal-fired steam plant it takes some time after the burners are lit for steam to start to be produced and put to the turbines, in a CCGT, some power will come from the gas turbine within about 15 minutes from startup. In industry parlance, the gas turbine is then 'synchronized'. It will take somewhat longer for the HRSG to get hot enough to produce steam; but both the gas and steam turbines of a CCGT can be up to full power within about an hour. Efforts to cut this time can damage the HRSG; but, on balance, CCGTs are probably better at meeting the bulk of the daytime load than steam plant.

Against this ability to start up quickly, the biggest shortcoming of CCGTs is the drop in efficiency at low loads, the cause being the drop in turbine temperature as power is reduced. Temperatures can be maintained down 80 per cent of the design rating by partially closing the inlet guide vanes of the compressor, thus reducing the flow of air along with the fuel supply. As a result, although less fuel is being burned, turbine inlet temperatures are maintained. Because of this characteristic, some operators have the view that although CCGTs are good for meeting the more stable part of the daytime peak, as well as being very good for base-load operation, they are not so efficient at compensating for highly variable loads. Such views are, of course, often coloured by individual experience with specific designs.

A new problem, related to the demand for power and its availability from CCGTs, is the increased air-conditioning load in the UK, which goes up on hot days. Unfortunately, the power output of CCGTs tends to fall at these times. High air temperatures reduce air density. Less fuel can be burned in the gas turbine, and the physical mass of air through the gas turbines drops away – so output declines. A drop-off in power can be overcome by burning extra fuel; but this can lead to turbine inlet temperature limits being exceeded and will result in some reduction in blade life. Palliatives include cooling of the inlet air; but this may not be economic if plants are only operating part-time.

EFFECTS ON PLANT COMPONENTS AND RELIABILITY

'Two-shifting'

Plants that have to 'two-shift' (i.e. be operated in intermittent rather than continuous mode) have to be kept in very good condition. Maintenance that is neglected risks unscheduled shutdowns. Even if this does not cause serious problems with the grid, it is likely to result in substantial financial penalties. Unfortunately, two-shifting is extremely damaging: rule-of-thumb estimates suggest that each start-up and shutdown is equivalent to about 20 hours of normal operation. The basic cause of the damage is the temperature gradients that develop and change as components heat up to normal operating conditions. But there are other factors that decrease reliability and increase operating costs. Many of these were discussed at a symposium on this subject, which covered the overall problem of two-shifting, rather than focusing simply on the metallurgical considerations (Shibli et al, 2001).

At this symposium, many of the presentations highlighted design problems in plants, most of which can be eliminated with good design and improved operating practices. Nevertheless, some problems are bound to increase as operating temperatures rise, as they will need to in more advanced power plant. These problems are of a metallurgical nature, being related to the stresses that result through the formation of temperature gradients.

The components that are operating at the highest temperatures suffer most from the thermal stresses. In a CCGT plant, the gas turbine, turbine blades and blade coatings tend to suffer from cracking; but the combustor cans are also susceptible. Fortunately, these components, although extremely expensive (typically resulting in repair bills in the millions of Euros range), are small enough to be replaced without too much difficulty. It is the bigger components, such as the high-temperature pipework and valves in steam plant, or the heat exchangers in an HRSG, that are more of a problem. Cracks will need to be cut out and filled in with weld metal. Some components will need to be replaced, also by welding. This is a major logistical issue and it is not always certain that the repairs will give the same life as the original equipment.

The reliability of HRSGs in CCGT plants under two shift conditions is giving considerable cause for concern. There are many potential problems; but just a few will be mentioned. On start-up, the gas turbine has to be brought online first, the result being that hot combustion gas flows through the HRSG duct for 10 to 20 minutes before much steam can be generated. Figure 6.3, adapted from Dooley et al (2003), shows that even when the gas turbine is carefully controlled, the temperature of the first row of superheater tubes in the HRSG climbs very rapidly once the gas turbine is fired up, increasing by over 300° C within five minutes of the gas turbine being lit. The temperature of the evaporator section of the HRSG shows a rather slower rate of rise. It is full of water and is located some way further back in the HRSG duct. The flat part of the evaporator curve is when the water is beginning to boil. As more steam is produced, pressure is increased and the evaporator temperature begins to move upwards again. What is not apparent from these curves is that individual tubes in both the evaporator and superheater heat up at different rates. The difference in temperature from tube to tube will result in some tubes being subject to tension, others to compression.

It follows that only when there is a good volume of steam flowing through the HRSG will the gas turbine be allowed to run at full output, and this does limit the ability of the CCGT to respond as quickly as it might to the demand for power. Although some power can be delivered to the grid reasonably quickly since the gas turbine can be synchronized with the grid within about 20 minutes of the start-up sequence being initiated, CCGTs are not as responsive as aero-derived machines.

Notice, in Figure 6.3, the long flat parts of the temperature curves during the first few minutes when everything is at about 30° C. This is the 'purge cycle', when the gas turbine is being motored up to its self-sustaining speed and pure air is being blown through the duct to flush out explosive gases. This purge is required on every start-up. It can be very damaging to the HRSG, especially if a restart is needed after a plant trip, when the HRSG is hot. These temperature changes can be damaging to the interior of the duct itself since it

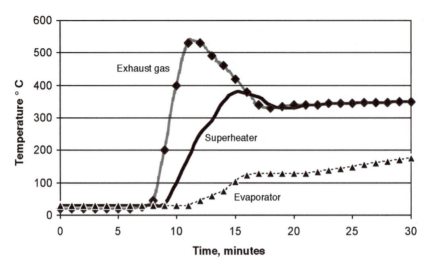

Source: adapted from Dooley et al (2003)

Figure 6.3 *Temperature changes in the front rows of a heat recovery steam generator superheater and evaporator*

has to respond to very rapid changes in the exhaust gas temperature. Unfortunately, trips are more likely during start-up than at any other time, and although the duct has just been swept clear of explosive gases, the purge sequence has to be gone through again.

Something similar happens to the HRSG during a normal shutdown. Initially, the gas turbine exhaust temperature will slowly drop as power is reduced. However, at below the self-sustaining speed, the fuel supply is cut and the temperature will drop more rapidly. Since there is still much steam in the HRSG at this time, condensation will occur. This phenomenon has caused considerable problems with some designs of HRSG since the condensed water will not necessarily drain away uniformly across the tube bank, once again inducing tube-to-tube temperature differences (see Figure 6.4). The situation is particularly critical if the plant is restarted after a trip, when some tubes contain water, while other tubes contain nothing but steam.

Damage to the superheater caused by steam condensation also occurs in coal-fired steam plant or a CCGT. Here, condensate can collect in the bottom of the 'U' tube loops of platen type, as shown in Figure 6.5. When the plant is restarted, relatively cold condensate can be carried over to the hotter parts of the equipment, where the sudden changes in temperatures can induce very high stresses. In very bad cases, the presence of condensate in a bottom loop will prevent steam flow through the tube. The result is that sections of the tube will run at over 800° C and literally burst open within a few minutes.

Temperature cycling problems can also exist with steam turbines. During normal operation, there is a temperature drop of about 150° C to 200° C from the inlet to the outlet of the turbine, as well as a temperature difference from

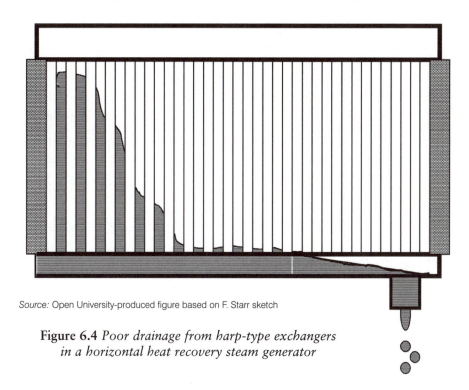

Source: Open University-produced figure based on F. Starr sketch

Figure 6.4 *Poor drainage from harp-type exchangers in a horizontal heat recovery steam generator*

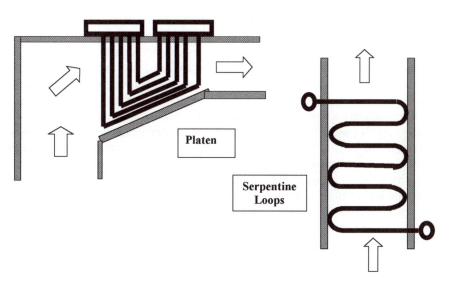

Note: Arrows show direction of flue or exhaust gas flow. Loop-type designs are common in vertical HRSGs and allow more satisfactory condensate drainage.
Source: Open University-produced figure based on F. Starr sketch

Figure 6.5 *Types of superheaters*

the outside to the inside of the turbine rotor. Thermal stresses will result. During weekend shutdown, the temperature gradients disappear, and this periodic change in the stress level can induce cracking.

These are just a few of the effects of two-shifting, which eventually will result in increased maintenance. The need for such repairs can catch operators and accountants by surprise. Initially, two-shifting does not have much of an impact on maintenance or plant reliability. Some reports suggest that it will take at least three years before the adverse effects become apparent. Hence, statements about plants being able to cope with intermittency without any difficulties need to be taken with a pinch of salt.

INTERMITTENCY AND POWER PLANTS OF THE FUTURE

Post-combustion capture units

In a strictly operational sense, there is no real problem in getting fossil fuel-generating plant to cope with intermittency. But what about the fossil plants that will be coming into service after 2015, particularly those that could be similar to those of today, but incorporating equipment to capture CO_2 from the flue gases? These are known as post-combustion CO_2 capture systems, in contrast to pre-combustion designs, based on gasification, which will be described in the subsequent section.

For coal-based plant, post-combustion designs will utilize the same type of furnace and steam system as those currently in use, but with the CO_2 being captured from the flue gas. The CO_2 can be absorbed using amine-based solvents that require steam for their regeneration. Another approach is that of the 'oxy-fuel process' in which oxygen is used for combustion of the coal rather than air, the aim being to increase the concentration of CO_2 in the flue gas. This permits carbon dioxide to be separated and captured using a simple compression process. But whether the CO_2 capture plant is of the absorption or of the oxy-fuel type, it is difficult to see how these plants can be two-shifted very easily. Shutting them down for days at time and then expecting them to operate at full power within a few hours is probably impossible. Part-load operation will be a technical possibility; but the economics need to be fully explored since the steam requirements of amine-type CO_2 capture and the electricity demand for an oxygen production unit will not necessarily decrease in proportion to plant output. The assumption, as with nuclear generation, appears to be that carbon capture-type plants will be run as base-load units; but this, to the author, seems not to be the safest of judgements, given that the life of power plants can be 30 years or more.

Post-combustion capture might also be applied to natural gas-fired CCGTs. In some respects, the operation will be easier than with coal-fired units as the amount of CO_2 produced per megawatt of electricity is much less. On the other hand, the percentage of CO_2 in the exhaust gas is lower than in coal-fired steam plant. A pragmatic approach might be to only absorb a proportion of the CO_2 that is being produced in CCGTs.

Turning now to the changes in the plants themselves, the real challenge is for coal-fired steam plant. The high level of carbon in the fuel points to the need to push efficiencies as high as possible to minimize the amount of CO_2 being produced. The target is an efficiency (before CO_2 capture) of over 50 per cent. This implies steam at 700° C and 350 to 400 bar pressure.

A new class of alloys will be needed for this duty. Even so, to contain the high pressure steam, the wall thicknesses of tubing, pipework, turbine casings and valves will have to be much heavier than current designs. More severe temperature gradients are likely to develop, resulting in higher stresses. In addition, some of these new materials, which are essentially sophisticated stainless steels, have a lower thermal conductivity and higher coefficient of thermal expansion than the materials in use at the present time. These factors, even discounting the increased wall thicknesses, will make equipment constructed from these new materials more susceptible to distortion and cracking when subjected to temperature changes.

The situation for CCGT plants does seem easier. In the author's view, the prospects for more radical approaches to gas turbine design, as exemplified by the new reheat and inter-cooled gas turbines now coming into use, seem very good. The basic thermodynamic cycle of inter-cooled reheat designs is a distinct improvement over the Brayton thermodynamic cycle, which is the basis of virtually every other form of gas turbine. In principle, this more advanced gas turbine cycle permits efficiency improvements without increases of turbine inlet temperature. The concentration of CO_2 in the flue gas will be higher than in more conventional designs, easing the capture problem.

Integrated gasification combined cycles in a renewable scenario

This analysis suggests that, because of the incorporation of CO_2 capture processes, the need to run at higher temperatures, and the temperature changes resulting from intermittency, current designs of plant may not be ideal. The need to find a more promising alternative is more pressing for coal. One such option is a modification of the Integrated Gasification Combined Cycle (IGCC), a gasification-based process that can easily be adapted to capture carbon dioxide, while at the same time generating hydrogen as a fuel gas.

The IGCC is a collective name for a variety of processes in which coal is gasified to produce a fuel gas, and which, after purification, is burned in the gas turbine in a conventional CCGT, built as part of the gasifier complex. Because the gasification process produces a large amount of waste heat, which can only be used for raising steam, the steam systems in the combined-cycle HRSG and gasification system are 'integrated' – hence the appellation IGCC.

IGCCs intended to capture carbon dioxide and produce hydrogen are likely to use gasifiers of the entrained flow type. Here, oxygen plus steam or water is reacted with coal at about 1300° C to 1600° C to produce a raw 'syngas' gas (which mainly consists of carbon monoxide and hydrogen, and is often used to synthesize chemicals, hence the name). After purification of the syngas to remove sulphur compounds, the mixture of CO and H_2 is used in a conventional IGCC as the fuel gas for the gas turbine. But to raise the hydrogen level and to enable the carbon in the coal to be captured as CO_2, the CO in the syngas is catalytically reacted with steam in a shift converter, the reaction being:

$$CO + H_2O \longrightarrow CO_2 + H_2 . \qquad [1]$$

After the 'shift reaction', the CO_2 would be absorbed using alkaline solutions, compressed and sent to a geological storage site. The hydrogen that remains could be used as fuel gas in the CCGT section of the plant. The overall reaction to produce hydrogen using coal as a fuel can be represented as:

$$C + 2H_2O \longrightarrow CO_2 + 2H_2 . \qquad [2]$$

The problem with IGCCs is that they not at all suitable as plants that have to be started up and shut down frequently. The IGCC gasifier and process train contains potentially explosive gases, and the necessary precautions on start-up could be lengthy and complex. Furthermore, an IGCC needs a great deal of ancillary plant for it to operate. These comprise units for removing hydrogen sulphide (H_2S) from the gas stream and oxidizing this gas to produce sulphur, a cryogenic air separation unit for producing oxygen, and the CO_2 capture plant itself. The conclusion would appear to be that a CO_2 capture-type IGCC has to be a base-load-generating system.

This is would be true if the only output from the plant was to be electricity. But by the time such plants are required, a hydrogen economy may be well developed in some countries, with a network of hydrogen pipelines. As a result, the Institute for Energy has proposed that at times when there is no demand for electricity, the hydrogen that the IGCC is producing should be diverted to the pipeline network. In this manner, the gasifier section of the IGCC could be kept at full output at all times.

Although fully supportive of this approach, it is the view of the author that it might be more practical for the UK to adopt a compromise situation, also based on the IGCC. This is recognizing that the UK economy is very dependent upon natural gas, accounting for over 70 per cent of the non-transport energy use. In keeping with this, the UK has a vast natural gas network, all of which has been renovated and expanded over the past 30 years. In this proposal, the purified syngas would be used to produce methane as a substitute natural gas (SNG). The methane from the gasifier stream could either be used as fuel gas in the gas turbine or, when electricity is not needed, could be put into the natural gas system as SNG. This should be a very attractive option for the UK as the SNG will supplement its declining gas reserves (see Figure 6.6).

To make SNG, a modified syngas mixture containing hydrogen, carbon monoxide and carbon dioxide are reacted together to produce methane:

$$CO + 3H_2 \longrightarrow CH_4 + H_2O \qquad [3]$$

and:

$$CO_2 + 4H_2 \longrightarrow CH_4 + 2H_2O. \qquad [4]$$

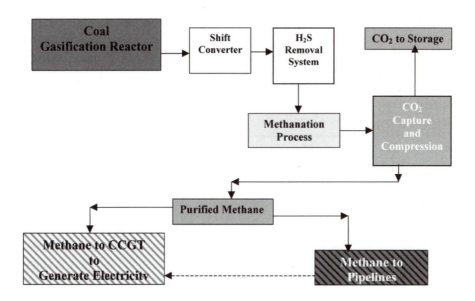

Figure 6.6 *Carbon capture coal to electricity and SNG*

But the overall reaction for the complete process of coal gasification, shift conversion and 'methanation' can be represented as:

$$2C + 2H_2O \longrightarrow CH_4 + CO_2. \tag{5}$$

It is apparent from Equations 2 and 5 that the main argument against an IGCC–electricity–methane process is that the rate of carbon capture is only half that when such a process makes hydrogen. On the other hand, such a system makes use of the existing UK natural gas infrastructure, which, of course, comprises not only the pipeline and distribution network, but also the Rough storage field in the North Sea, the liquid natural gas (LNG) storage sites, and also, at the consumer level, the burners (which will only work using a methane rich gas), in central heating systems, gas cookers, industrial furnaces, and natural gas-fired CCGT plants.

Whether the fuel gas is hydrogen or methane, the combined-cycle section of the IGCC would still be subject to start and stop operation; but it would have considerable advantages over the natural gas-fired CCGTs of today. The HRSG section of the plant could be kept hot by bleeding off steam from the gasifier steam system. This will eliminate much of the temperature cycling to which a normal HRSG is subject. It also gives the option of an extremely fast start-up since the HRSG is kept hot.

There are other advantages. In principle, if liquid oxygen were stored, it should be possible to run the cryogenic plant at a reduced output, saving some power at times when there was a high demand for electricity. This would help to overcome one of the problems of CCGTs – that is, reduced power output on hot days. More important is the prospect of constructing the gasifier to be undersized in relation to the declared electrical output from the plant. The CCGT would utilize some of the hydrogen or methane that was stored in the pipeline network. This ability is shown schematically in Figure 6.6. Both of these ideas would have a significant effect on the capital costs.

CONCLUSIONS

The author supports the opinion that the effect of intermittency on the present set of UK generating plants will not cause operating problems and is not substantially different from what happens when steam or CCGT plants are designated as two-shift units. The main implications are that irregular operation will lead to increased maintenance costs and unreliability. These factors are well recognized within the power generation sector, and their impact can be minimized by improved detail design and more sympathetic operating procedures. The real difficulty seems to be with future designs of fossil fuel-generating plant.

But this chapter has also indicated that too much may be taken for granted, particularly with advanced coal-fired steam plant. Even if fossil-fuelled plants of the future are not of the carbon-capture type, efforts to improve plant efficiency through raising temperatures and pressures will require heavier section components made of materials that are inherently less tolerant of stop–start operation.

The addition of systems to capture CO_2 will also make any type of plant less able to meet the demands of intermittent operation. The standard type of IGCC plant shares these shortcomings; but it is possible to design a modified form of carbon-capture IGCC that can switch its output from electricity to hydrogen or SNG as demand for electrical energy falls. This type of generating plant offers high efficiency and, in the coming age of renewables, will have the ability to respond to short time variations and to provide electricity extremely quickly.

REFERENCES

Babcock and Wilcox (2004) *Steam: Its Production and Use*, 41st edn, UK Government Department of Trade and Industry (DTI), London

DTI (Department of Trade and Industry) (1999) *Supercritical Power Cycles: A Technology Status Report*, DTI, London

DTI (2004) *Heat Recovery Steam Generators*, DTI, London

DTI (2006a) *The Energy Challenge,* UK DTI Report on the Energy Review, URN no 6/1576X, DTI, London

DTI (2006b) *Reducing the Cost of System Intermittency by Demand Side Control Measures*, DTI, London

DTI (2006c) *Advanced Power Plant Using High Efficiency Boiler/Turbine*, DTI, London

Dooley, B., Paterson, S. and Pearson, M. (2003) *HRSG Dependability*, ETD Ltd, London

Environmental Audit Committee (2006) *Keeping the Lights On: Nuclear, Renewables and Climate Change*, UK House of Commons Environmental Audit Committee, Sixth report of Session 2005–06, HC-584-I, London

European Parliament (2004) *Directive 2004/8/EC*, Brussels

Pehnt, M. et al (2005) *Micro Cogeneration: Towards a Decentralized System*, Springer Verlag, Berlin

Shibli, I. A., Starr, F., Viswanathan, R. and Gray, D. (2001) *Cyclic Operation of Power Plant: Technical, Operational and Cost Issues*, ETD Ltd, Ashtead

Starr, F., Tzimas, E. and Peteves, S. D. (2005) *Flexibility in the Production of Hydrogen and Electricity from Fossil Fuel Plants*, IHEC, Istanbul, Turkey

Starr, F., Tzimas, E. and Peteves, S. D. (2006) *IGCC and Steam Reforming Plants for the Production of Hydrogen and Electricity: The Development Issues*, EUR Report, EUR No 22340 DG-JRC Institute for Energy-Petten, The Netherlands

Quaschning, V. (2001) *Simulationserebissne für die Regenerative Erzeugung in Jahr 2050*, Unpublished report

The views expressed in this chapter are those of the author and not necessarily those of the European Commission.

7

The Potential Contribution of Emergency Diesel Standby Generators in Dealing with the Variability of Renewable Energy Sources

David Andrews

INTRODUCTION: WESSEX WATER

Wessex Water is one of ten water and sewerage companies in England and Wales, covering Somerset, Dorset, Wiltshire and parts of Avon. Energy is one of the company's largest operational costs: average electrical power use is about 27 megawatts (MW). The company has about 8MW of biogas combined heat and power (CHP) generation capacity, of which 4.5MW is continuously operating, provided by spark-ignited gas engines fuelled by digester gas. It also has some 550 emergency standby diesel engines, totalling 110MW of capacity, whose primary function is to power essential services such as sewage works and water supply works during power failures, which happen, on average, a few hours each year. Of this number, about 33 units, totalling 18MW, are also used commercially in a number of non-emergency ways that we call 'load management'. This includes routinely feeding power into the local electricity distribution system and, ultimately, the UK National Grid. These generators currently have a four-minute automatic start-up and paralleling capability, and are currently being modified to enable start-up in less than one minute. The units are quite small, ranging from 0.24MW to 1MW, and are used by the National Grid on a regular 'call-off' basis to supplement its arrangements with power station owners.

THE NATIONAL GRID TRANSCO FREQUENCY SERVICE

National Grid Transco (NGT), which operates the national grid and controls the operations of power stations in England and Wales, has a number of partners known as NGT Frequency Service participants. These are large power users, such as steel works or cold stores, who enter into a contract to be paid to be disconnected from power supplies whenever grid frequency starts to fall. For example, a very large steel melting furnace, which may take a day to heat up using an electric arc or induction heater, is not adversely affected if the process is delayed by 20 minutes; but this can obviously help the grid enormously if a sudden power demand is being made on the grid. The same applies to a large cold store where interruption in cooling for 20 minutes is unimportant.

This instant switch-off is achieved using a relay provided by NGT, linked by telemetry to the NGT control centre and mounted on the incoming power supply switch gear. It is set to detect the falling frequency that can occur when a large power station fails suddenly, and opens the circuit breaker to a demand centre. The relay can be remotely monitored by NGT, who can control the exact frequency at which the relay disconnects the load and can monitor whether the relay is armed or not, whether the customer has temporarily exercised his right to override the relay, etc. Frequency Service participants are contracted to stay off for up to 20 minutes. They receive a fee of the order of several thousand UK pounds per megawatt of capacity per year.

THE NATIONAL GRID TRANSCO RESERVE SERVICE

Operating closely with NGT Frequency Service is the NGT Reserve Service. Participants include owners of small diesel (in the range 0.25–5MW) or open-cycle gas turbine generators (in the range of 25MW to 100MW) who are paid to start up and connect to the grid within 20 minutes at the same time as Frequency Service customers are called upon to disconnect. Participants must be reliable and able to stay connected and running for an hour or so. Substantial fees can be earned simply for making generating capacity available under complex contracts that specify levels of reliability, response times, frequency of use and so on.

RESERVE AND STANDBY GENERATING CAPACITY ON THE UK NATIONAL GRID

On the UK National Grid system there is approximately 1.5GW of 'spinning reserve' – typically, this takes the form of a large power station that is paid to produce at less than its full output. Such a station might have four generating sets each of 660MW, giving a total output of 2.64GW, but might only be operating at 2GW with the steam boiler full but the steam valve not fully open. On request from the National Grid control centre, this valve can open and deliver an extra 640MW in 20 to 30 seconds. This requires the boiler air fans and the coal feeders to increase output accordingly. The greater the total load on the system, and the greater the expectation of large fluctuations (e.g. at the end of popular TV programmes), the larger the proportion of spinning reserve set by the NGT.

It is worth noting that the cost of such spinning reserve is not high, as is often erroneously stated. The efficiency of a plant might change from, say, 37 to 36.5 per cent if the output of the set is dropped from 660MW to 500MW (i.e. 160MW spinning reserve). The fuel penalty involved (about 1.5 per cent) is tiny compared to the total amount of fuel passing through the power station.

The NGT also pays to have up to 8.5GW of additional capacity available, but not running (known as 'warming' or 'hot standby' capacity), which can take as little as two hours or, in some cases, half an hour to bring online. Generally, there will be more of such hot standby capacity the greater the expected disturbance on the system. The cost of fuel required to keep such plant warm is tiny in comparison with the amount of fuel used to generate power.

A similar amount (8GW to 10GW) of plant is operable from cold in about 12 hours for coal plant, and around 2 hours for gas-fired plant.

The pumped storage schemes at Dinorwig and Ffestiniog can offer 2GW of power within 15 seconds, and the Cross Channel high-voltage link can bring in up to 2GW of power from France. In addition, as described above, the NGT can call on its Frequency Service and Reserve Service participants.

Consider what happens if a typical large 660MW turbine generator set suddenly 'trips'. This can happen for all sorts of reasons – a coal crusher might break down, boiler tubes might fail, an alternator might start to overheat, or insulation might fail on the alternator. Because the grid has suddenly lost 660MW, which on a typical day might be 1.3 per cent of total output, then due to the immediate imbalance between supply and demand, grid frequency immediately starts to drop from the standard 50 hertz (Hz).

As soon as this happens, the under-frequency relays on Frequency Service customers begin to trip off their loads as frequency falls, ultimately shedding loads equal to 660MW. These relays are set at a random range of frequencies between 48.5Hz and 49.5Hz, so the 660MW of generation that has been lost is not instantly matched by these relays shedding 660MW of load simultaneously. Instead, this happens progressively as the frequency drops until exactly enough is shed to exactly match the remaining power station capacity. This will then stabilize the frequency at a lower level – perhaps 49.3Hz. All this happens in a few seconds.

Frequency Service participants are only contracted to have their shed loads off for up to 20 minutes; as a result, the NGT control room issues start-up signals to enough of its Reserve Service participants to enable up to 660MW to become available within 20 minutes. NGT control monitors the situation, and if sufficient Reserve Service capacity does not come on, it can order more until it has exactly matched the load that the Frequency Service relays have shed.

When sufficient Reserve Service capacity has become available in less than 20 minutes, the Frequency Service loads (steel furnaces, cold stores, etc.) are re-connected by the NGT – gradually, so as not to destabilize the system. The Frequency Service relays are then re-armed by NGT.

Up to an hour or so later, the output of the Reserve Service diesels and gas turbines (which are nearly all in private hands, and not professional power generators) will have been augmented and then replaced with increased levels of generation from large gas- or coal-fired power stations, such as those on spinning reserve. These together will have driven the frequency back to its correct level, at close to 50Hz. The diesels can then be stood down, ready for the next emergency.

At the same time, new levels of spinning reserve will have been created, which might have been stations on hot standby now switched to running. Increased levels of hot standby capacity will also be called for.

At present, the largest sources of intermittency on the National Grid are the power stations themselves. For example, whenever the UK's largest nuclear power station, Sizewell B, is operating, its entire output is capable of dropping to zero at any time, with little or no warning. Its capacity is 1.2GW, around 2 per cent of the National Grid maximum demand. Yet, the NGT readily copes with such failures by using the methods outlined above.

The kind of intermittency that a very high proportion of wind power plant on the grid would introduce is much less than the intermittency already there due to large conventional power stations. Even in the most extreme case, the simultaneous change in output of all wind turbines in the UK would take many minutes to achieve the instantaneous and unpredictable change in output caused when Sizewell B trips.

Furthermore, the most reliable form of wind forecasting is to simply look at the total output of the wind turbines. There is a high probability that the power they are producing at any given time will be similar to that produced one hour later. As this prediction 'window' is decreased – to 20 minutes, 10 minutes or 5 minutes – the difference in predicted total national wind power output becomes less and less, and even at five minutes, there is ample time to raise or lower spinning reserve accordingly. If the 5-minute estimates are wrong, then the Frequency Service and Reserve Service diesels will have the resilience to cope with it.

'Triads': A revenue-earning opportunity

So-called 'triad' periods provide a further revenue-earning opportunity, separate from the Reserve Service. Triad periods are the three half-hour periods of maximum electricity demand during winter. Diesel generators can earn substantial sums by reducing a site's peak demand during these peak periods.

National Grid Transco is funded principally by means of a capacity charge levied on the energy suppliers, who then pass it on to their customers in a more or less transparent way.

The charge is calculated in retrospect by the NGT looking back over each of the 17,520 half hours in a year and locating the three half hours, separated by at least ten days, of total NGT system maximum demand, which at peak might approach 60GW. Having identified these triad half hours, it then charges each of the energy supply companies according to their average peak loads on the National Grid system during those three periods. For example, at the western extremity of the system, the Western Power Distribution (WPD) area in the south-west, the total annual transmission cost is about UK£21,000 per megawatt per year. So, if an energy supply company can cut its load by 1MW or start a 1MW diesel during triad periods, it can save UK£21,000, compared with a fuel cost of perhaps only UK£150.

However, it is not easy to predict exactly when the triads are going to occur. Therefore, in order to ensure 'triad capture', Wessex Water starts its generators about 30 times per year for about one hour, expending approximately UK£3000 on fuel. Since triads always occur at times of high power prices, savings are also obtained from avoiding the purchase of power during such periods: this could cost about UK£3000, which more or less offsets the cost of the diesel fuel used.

For advice on the likelihood of a triad occurring, Wessex Water pays for a triad forecasting service, which typically arrives at 11.00 am in preparation for a 5.30 pm run later that day. Triads always occur at about 5.30 pm on winter weekdays, except on Friday.

Perhaps surprisingly, Wessex Water also has contracts with generating and energy supply companies, who also pay to operate its diesels remotely from time to time, for balancing purposes and when they are short of capacity.

Wessex Water clearly cannot be running for triad period supply and for other generating companies when it has taken a capacity payment from the NGT to keep its generators available. However, the company's contracts enable it to declare its generators unavailable during the anticipated triad periods – so the NGT will call upon another generator. Wessex Water declares the status of its generators automatically in real time to third parties so that they also know when they can or cannot be available.

OTHER BENEFITS: TESTING DIESELS OFF LOAD

Diesels must be run regularly at least once a month, and preferably once a week, to ensure they will work when called on unexpectedly in an emergency power failure. Failure to run diesel generators regularly means they are very unlikely to start in an emergency, usually due to a variety of simple failures – most commonly, flat batteries, contaminated fuel or corroded contacts. Even if they do start, they are likely to stop after a short while – usually overheating due to failures of the cooling systems.

There is a general tendency to assume that it is preferable to run diesels off load: it is assumed that this is less harmful than wearing them out by running them at full load. However, this ignores the fact that diesel generators are designed to run at their stated rating.

In fact, testing diesels off load is extremely harmful and can very quickly ruin an engine – in as little as only 50 hours of accumulated running. This is because under-loading causes a series of damaging interlocking events. Initially, this involves low cylinder pressures and consequent poor sealing of piston rings – these rely on the gas pressure to force them against the oil film on the bores to form the seal. Low initial pressure also causes poor combustion and resultant low combustion pressures and temperatures. This poor combustion leads to soot formation and unburned fuel residues, which clog and gum piston rings yet further, causing an additional drop in sealing efficiency and exacerbating the initial low pressure.

Hard carbon also forms from poor combustion. This is highly abrasive and scrapes the honing marks on the bores, leading to bore polishing, which then results in increased oil consumption ('blue-smoking') and yet further loss of pressure since the oil film trapped in the honing marks maintains the piston seal and pressures.

Unburned fuel leaks past the piston rings and contaminates the lubricating oil. At the same time, the injectors are being clogged with soot, causing further deterioration in combustion and black smoking. This cycle of degradation means that the engine soon becomes irreversibly damaged, may not start at all and will no longer be able to reach full power when required.

Under-loaded running inevitably causes not only white smoke from unburned fuel due to the engine's failure to heat up rapidly, but over time, as the engine is destroyed, it is joined by the blue smoke of burned lubricating oil leaking past the damaged piston rings and the black smoke caused by the damaged injectors. This pollution is unacceptable to the authorities and any neighbours.

With a fully loaded diesel, there is only a very short puff of white smoke that rapidly disappears once the diesels warm up in a matter of seconds.

The internationally agreed definitions of the power rating levels for diesel engines are:

- standby: short-term use, only for ten hours per year;
- prime power: where the generator is the sole source of power for an off-grid site, such as a mining camp or construction site, and demand is continuously varying;
- continuous: output that can be maintained 8760 hours per year.

Typically, if the standby rating is 1000 kilowatts (kW), then the prime power rating would be 850kW and the continuous rating 800kW.

Wessex Water sets are sized initially on the standby rating for emergency use, but are run in load management mode at the continuous rating level, which is about 80 per cent of the standby rating.

A diesel engine can be tested on full load by connecting it to a load bank; but this usually means hiring a load bank and a specialist to physically connect it, which is an expensive operation. Alternatively, a dedicated load bank is sometimes provided; but this itself has a cost and is obviously a fuel waster.

The generator could, of course, be used to run the emergency load to which it is connected; but this usually means an undesirable break in supply unless short-term paralleling devices are fitted. Generally, the load connected to a generator is found to be only about one third of its maximum standby rating, so this can lead to long-term problems – although this is not nearly as bad as no-load running.

One often finds that major defects are pre-emptively identified by load management runs. For example, in a recent case at Wessex Water's Weymouth head works site, the generator caught fire due to a failed turbo oil seal; this would have occurred sooner or later, but it was greatly advantageous that it occurred during a load management run and not during an emergency run, and was therefore repaired before the next power outage.

Therefore, load management is the ideal way to prove diesels without destroying them because it gives a readily available full load test – and earns income.

In the UK, there is perhaps up to 20GW of emergency diesel capacity. With the right financial incentives and explanations of the benefits, a large proportion of this could be brought into Reserve Service and similar schemes. I would expect that in 20 years this practice and the associated technology will become standard.

CONCLUSIONS

Very small diesel generators already play a vital role in stabilizing the UK national grid, using only about 10 per cent of the available generators. Clearly the use of these unused sets could be expanded very economically to assist with huge expansions of wind power and the likely power supply fluctuations that would sometimes occur.

8

Demand Flexibility, Micro-Combined Heat and Power and the 'Informated' Grid

Bob Everett

INTRODUCTION

We are continually being told that energy efficiency is cheaper than energy supply. If this is so, then the first response to coping with a varying electricity supply source should be to turn off or delay a load, rather than start up yet another generator. The overall potential must be enormous. It can be judged by the drop-off in UK electricity demand that took place during the 1999 total eclipse of the sun (see Figure 8.1). Over the space of half an hour, demand dropped by 2.2 gigawatts (GW) as people found something better to do than work. Over the next half hour it rose by 3GW as they went back to work, no doubt after brewing a celebratory cup of tea or coffee.

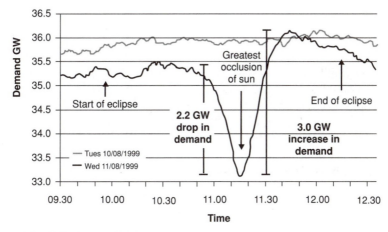

Source: National Grid Company (1999)

Figure 8.1 *UK National Grid demand during the total solar eclipse, 11 August 1999*

WHAT IS NEEDED

There are many ways in which demand-side management could allow a higher penetration of variable renewable energy electricity-generation sources; some of these are described in other chapters.

Overall flattening of day–night demand

There is considerable variability in electricity demand throughout the day, even without the introduction of variable renewable electricity supplies. It would thus pay to 'flatten' the day–night electricity demand curve by encouraging the use of off-peak electricity. This would probably require an increase in the use of electric heating. Given the availability of cheap North Sea gas, this has not been economically viable in the UK since the 1970s, but could become so again in the future. It would be beneficial for the economics of all forms of electricity generation, particularly those with high capital costs. In the past, this has meant nuclear power; but it could equally apply to wind and tidal power.

Frequency service to cope with failures of large power plants

As described in Chapter 7, hundreds of megawatts of backup power are necessary for a period of up to half an hour to cover the failure of a large power station. It would be useful if the demand-side role for this could be expanded.

Conventional short-term reserve service to cover wind prediction errors

As described in Chapter 5, weather forecasting can predict the wind speeds that a wind farm is likely to experience fairly accurately. What is not so accurate is the precise timing of *when* a weather front is likely to arrive. This can lead to large short-term prediction errors. The magnitude of the error could be many gigawatts, depending upon the amount of installed wind capacity and its geographical spread. The time scale of the error may be of the order of an hour or more, depending upon how far ahead the prediction is being made. Large amounts of local reserve service may be needed for balancing if grid connections between different wind farm areas are weak, or if 'local balancing' is chosen for operational or contractual reasons. Either way, it would be convenient if the demand side could supply a gigawatt or more of short-term flexibility.

Longer time scale reserve service

Longer-term flexibility will be needed to cope with 'the day with no wind'. The wind will, of course, eventually return as weather fronts track across the country. Typically, these may cross the UK every four or five days, so it would be useful to have loads that could be delayed for this amount of time. In the longer term, loads that could be delayed for up to six hours or so would be useful to cope with the gaps between the output of any future large tidal power schemes.

WHAT IS ON OFFER

Domestic refrigeration

Refrigerators are one type of load where the demand could be delayed. The idea of equipping refrigerators to give short-term frequency service has been promoted by companies such as Dynamic Demand (see www.dynamic-demand.co.uk). It has developed a controller that monitors the mains frequency and turns the fridge off if it falls below a preset limit. This concept has received official support in the 2006 Climate Change and Sustainable Energy Act (HMSO, 2006), which requires the Secretary of State to publish a report on the technology's potential to save greenhouse gas emissions.

While this seems basically a good idea, its target device – the domestic refrigerator – is undergoing transformation with European Union (EU) legislation requiring better insulated, more efficient designs. It has been estimated that annual UK electricity used by domestic 'cold' devices peaked at 17.5 terawatt hours (TWh) in 1999, falling to 15TWh in 2004 (DTI, 2006), due to the increasing proportion of new high-efficiency designs. This trend is likely to continue. It has been suggested that with new insulation technology, by 2050 this figure could have fallen to only 3.5TWh (Boardman et al, 2005). Even so, this would represent an almost continuous system load of 400MW that could be flexibly controlled. There is no reason why this control technology could not be used in other applications, such as dimming street lamps, which would add to this potential.

As refrigerators become better insulated, so the length of time that the supply could be delayed increases. One modern A++ rated domestic freezer is advertised as being able to keep food frozen for 64 hours after a mains interruption.

Commercial refrigeration

Supermarkets and their associated distribution depots have large cold stores. As with domestic refrigerators, the supply to these could be delayed for short periods in order to cope with the needs of the grid. A recent study suggested that these could offer about 300MW of short-term flexible demand within the UK (IPA Consulting et al, 2006). As with domestic fridges, this potential could fall in the longer term if higher overall insulation standards are brought in for such cold stores.

Large-scale water pumping

The study cited in IPA Consulting et al (2006) also looked at the possibilities for demand-side flexibility in the water industry, where large amounts of electricity are used for pumping. It concluded that the UK potential for delays of one to three hours was almost 300MW. The time delays could, of course, be increased given investment in suitable storage reservoirs.

Off-peak domestic electric heating

According to the Building Research Establishment (2006), some 1.3 million homes in Great Britain have electric storage central heating, using night-time electricity supplied at a cheap rate. Water heating is carried out by an immersion heater element, typically of 3 kilowatt (kW) rating, in an insulated water storage tank. Space heating uses storage heaters that can have a typical input rating of 10kW to 15kW per house.

The Economy 7 tariff guarantees seven hours of cheap-rate electricity at night; but the precise timing is controlled by the National Grid Company. The switch-over from 'day' to 'night' tariffs is carried out by radio-controlled time switches in each home. This technology (introduced during the early 1980s) uses an inaudible sub-carrier on the three long-wave Radio 4 transmitters. There are 15 channels available, enabling a progressive switching of the off-peak load. There is obviously a potential for flexibility in delaying this load, or even temporarily switching it off in the middle of the night. Having short-term flexible demand that can only be used at night may seem restrictive; but the load available is large.

Water heating is an activity that is likely to be carried out all year round. A total of 1.3 million 3kW immersion heaters represents a national load of nearly 4GW. In addition, there are likely to be many more off-peak immersion heaters in non-centrally heated homes.

Domestic micro-combined heat and power

Domestic-scale micro-combined heat and power (CHP) units using Stirling engines are now being marketed, and prototypes using fuel cells are being tested. The potential for this technology is enormous. A report by the Society of British Gas Industries (2003) suggests that these could be potentially fitted in 14 million UK homes as gas boiler replacements that also generate some electricity. If each of these were rated at 1kW electrical output, this would represent a total installed electrical generation capacity of 14GW. Current models are only designed to operate as substitute (electricity-generating) gas boilers. As such, they are controlled by the heat load of the house, rather than its electricity load. There is, thus, an assumption that the aggregate output of many hundreds or thousands of these will be a reasonable match to the average electricity demand of these homes.

There is the future possibility of modify the timing of the CHP system's operation, particularly if the system incorporated some heat storage, such as a conventional hot water cylinder. This would allow it to respond to the needs of the electricity grid as well as just the heat needs of the house. If domestic micro-CHP were to become a successful technology, with millions of installed systems, then the total national potential to make use of this flexibility could amount to several gigawatts.

This would, of course, require some remote information link to the micro-CHP system. Householders may wonder why their 'gas boiler' needs access to a telephone line or a radio link; but there is a precedent for this. During the early days of small-scale (circa 100kW) CHP in the 1980s, the technology gained a reputation for poor reliability. This situation was transformed by installing modem links to the equipment suppliers that allowed online monitoring and fault diagnosis. Servicing and rapid repairs could be carried out before problems became serious, expensive and embarrassing. The installation of similar links to domestic micro-CHP units could prove equally useful. Online monitoring could give increased reliability and consumer confidence, and allow the designs to be pushed to higher efficiencies.

Once installed, the information link could be used for many other purposes, including some form of remote scheduling of generation, making domestic micro-CHP a complementary generation technology to wind power or other variable energy sources.

METERING AND THE FUTURE

A major bottleneck exists at the metering interface between the small consumer and the electricity supplier. This has not yet caught up with the information age. The basic 'spinning disk' meter was introduced at the end of the 19th century, and attitudes towards it are largely unchanged. Meters are regularly only read once a year and consumers (even quite large ones) are presented with estimated bills for the rest of the year. Thus, most consumers are unaware of the finer details of their electricity use and are completely unable to react to the changing needs of the electricity grid.

The increased use of micro-generation, such as photovoltaic panels and domestic micro-CHP, brings with it the need for two-way metering and billing for both imported and exported electricity. On-peak/off-peak meters are already remotely radio controlled. It is only a short step to foresee a full two-way flow of information between generators and consumers.

It has been suggested that such an 'informated grid' could give rise to whole new decentralized markets for electricity (Awerbuch, 2004). Thus, if a load were deemed to be relatively unimportant, it would not be supplied until the electricity price had fallen to an appropriate level or a certain amount of time had elapsed. Domestic micro-CHP plant could be remotely scheduled. At present, the capital cost of 'smart meters' appears to be a stumbling block. There is a perverse logic to this – the cost of a 'smart meter' may seem high compared to the average domestic electricity bill. On the other hand, if these meters allow the flexible use of limited energy supplies, it may be money well spent.

This leaves some unanswered questions:

• Who will do the controlling?
• How will they interface with the current market structure?

As usual, further research will be needed.

REFERENCES

Awerbuch, S. (2004) *Restructuring our Electricity Networks to Promote Decarbonisation*, Tyndall Centre Working Paper 49, www.tyndall.ac.uk/publications/pub_list_2004.shtml, accessed 10 November 2006

Boardman, B., Darby, S., Killip, G., Hinnells, M., Jardine, C. N., Palmer, J. and Sinden, G. (2005) *40% House*, Environmental Change Institute, Oxford, www.eci.ox.ac.uk/research/energy/40house.php, accessed 10 November 2006

Building Research Establishment (2006) *Domestic Energy Fact File*, BRE, Watford, www.bre.co.uk, accessed 10 November 2006

DTI (Department of Trade and Industry) (2006) *Energy Sector Indicators*, DTI, London, www.dti.gov.uk/energy, accessed 10 November 2006

HMSO (Her Majesty's Stationery Office) (2006) *Climate Change and Sustainable Energy Act 2006*, HMSO, London, www.parliament.uk, accessed 10 November 2006

IPA Consulting, Econnect Ltd and Martin Energy (2006) *Reducing the Cost of System Intermittency Using Demand Side Control Measures*, www.dti.gov.uk/files/file33122.pdf, accessed 10 November 2006

Society of British Gas Industries (2003) *Micro-CHP: Delivering a Low Carbon Future*, Leamington Spa, www.sbgi.org.uk, accessed 10 November 2006

9

A Renewable Electricity System for the UK

Mark Barrett

INTRODUCTION

The world faces a combined environmental and energy challenge: global warming and the depletion of fossil fuels.

Emissions of greenhouse gases, principally carbon dioxide (CO_2), are causing global warming, which will make conditions difficult for human beings and ecosystems. The main source of anthropogenic CO_2 is the burning of fossil fuels. Ultimately, a reduction in the global emission of anthropogenic CO_2 of more than 60 per cent is required so that the rate of climate change and the final global temperature increase are not so extreme that the impacts on humanity and ecosystems are catastrophic. Under the Kyoto Protocol, the UK has a target of a 12.5 per cent reduction in 1990 greenhouse gas emissions by 2010. Beyond this, the current UK government has aspirational target reductions of 20 per cent by 2010 and 50 per cent by 2050.

At the same time, industrialized high-consuming societies rely heavily on the supply of fossil fuels (gas, oil and coal) for the majority of essential and leisure services. The remaining reserves of gas and oil will be severely depleted over the coming decades, and global competition for these fuels will intensify. UK production of oil and gas will decline rapidly, and providing services essential to current lifestyles with these fuels alone will not be possible in the medium term. Coal reserves will last much longer at current depletion rates; but coal emits more CO_2 than the other fossil fuels per unit of energy produced.

Demands for energy vary with social and economic activity, and some depend upon the weather. The demands for energy for heating, lighting, cooking, industrial processes, transport and so on vary hour by hour throughout the day, the week and across the seasons; demands also vary spatially according to human settlement patterns and economic activities. Fossil fuels exist in natural reserves or stores; they may be extracted as quickly as required, and they are easily transported and distributed to the point of use. They may be easily stored, even in mobile vehicles, ships and aircraft. The supply of energy from fossil fuels may thus be easily adjusted to meet demands as they vary in time and space.

Fortunately, the solutions to the problem of reducing CO_2 emission will also address the other problem: the exhaustion of finite fuels. The problems of controlling climate change and of replacing fossil fuels can be addressed by minimizing energy demand with energy efficiency, and by replacing finite fuels with renewable energy sources. Nuclear power is a low carbon energy source; but it relies on a finite source of energy – uranium – and it has severe safety, public acceptability and economic problems.

Renewable energy is, by definition, inexhaustible. Unfortunately, the various sources of renewable energy fluctuate widely over different time periods. For some sources (e.g. tidal power), the fluctuations are quite predictable; but for others, such as wind power, it is difficult to predict their intensities many hours or days into the future. Apart from biomass and, to an extent, hydro, renewable energy sources are not naturally stored to any degree, and artificial stores have therefore to be made as required. Renewable resources also vary spatially. The challenge, then, is find out how variable renewable energy sources and storage can be combined in order to meet energy demands reliably in the most cost-effective manner. This problem is particularly acute for electricity supply because electricity is expensive to store in significant quantities.

This chapter summarizes the results of a study based on an hourly, dynamic physical energy model of a future UK electricity supply system based almost entirely on renewable sources of electricity. It demonstrates the technical feasibility of a 95 per cent renewable electricity system, a finding that is of strategic importance to the UK.

Key findings of the study to date are:

- An electricity system is feasible in which 95 per cent of the energy is renewable; electricity can also be supplied reliably, hour by hour, over the whole year.
- Emissions of greenhouse gases and air pollutants from electricity generation can be virtually eliminated.
- The unit costs of electricity in the system are not excessive, compared to future finite-fuelled generation, and are not subject to the uncertain availability and price of imported finite fuels.
- The system is secure in the long term because it is largely based on indigenous energy sources and does not employ irreversible technologies that pose substantial risks.
- The renewable electricity can be used to substitute for some gas and liquid transport fuels.

The goal is to design a sustainable electricity service system that meets environmental objectives for global warming and air pollution, and that also reduces or eliminates several categories of risk associated with other electricity supply mixes.

The electricity service system proposed would reduce carbon dioxide emissions and provide secure energy supplies in a context of declining UK fossil fuel production and nuclear generation.

This electricity service system is one in which 95 per cent of electricity demand is met by renewable energy sources sited in the UK. The options exercised in the scenario to be described here include energy efficiency improvements, the large-scale introduction of renewable electricity-only sources, and biomass-fuelled combined heat and power (CHP). It is also assumed that fossil-fuelled generation is used for firm capacity beyond that provided by biomass CHP, that the trade link between the UK and France is reinforced, and that existing nuclear stations are not replaced. The systems modelled result in very low emissions of greenhouse gases and other atmospheric pollutants because fossil fuel use is small. This system is one that might be put in place over the next 40 years as nuclear and fossil generation decline.

It is argued that the system proposed here is technically and economically feasible and would meet environmental and energy security objectives. *Prima facie*, the system is more desirable than one based on finite energy sources, such as coal or nuclear power. It is therefore argued that such sustainable systems should be further developed and assessed before strategic decisions involving irreversible technologies are taken.

The remainder of the chapter describes:

- the overall energy scenario context, including sectors other than electricity;
- the components of the electricity system: demand, storage and generation; an optimized integrated system.

SCENARIO CONTEXT

The provision of electricity services should not be planned in isolation from the overall UK and, indeed, European energy systems. Energy planning should be integrated across all segments of demand and supply. If this is not done, the system may be technically dysfunctional or environmentally and economically suboptimal. Energy supply requirements are dependent upon the sizes of, and variations in, demands over time scales from minutes to years, and these rely on future social patterns and technologies. Some examples of integrated planning issues are as follows:

- CHP electricity generation depends upon the associated heat load, and this, in turn, depends upon insulation in buildings and how much heat is provided from other sources, such as solar energy.
- Electric vehicles will add to electricity demand, but reduce fossil liquid fuel consumption and add to electricity storage capacity, which aids renewable integration.
- Is it better to burn biomass in CHP plants and produce electricity for electric vehicles, or inefficiently convert it to biofuels for use in conventional internal combustion engines?

Scenario context: Dwelling space heating

As an example for context, the interaction between dwellings and energy supply is briefly explored. This is only to illustrate and emphasize that determining optimum supply mixes can only be properly done with detailed analysis extending across all segments of demand, and across supply systems other than electricity.

The implementation of space heat demand management (insulation and ventilation control) will change the amount and time variation of heat demand in the domestic sector. Figure 9.1 shows a possible profile over the coming decades for maximum demand management as building regulations and refurbishment take effect.

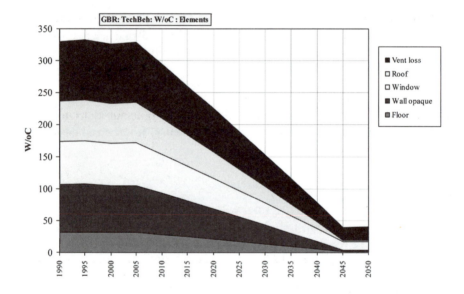

Note: The title *GBR: TechBeh: W/°C: Elements* means *Great Britain: Technical and Behavioural Scenario: Heat Loss in W/° C of Different Elements of a Building.* 'Vent loss' is ventilation loss.
Source: Society, Energy and Environment Scenarios (SEEScen) model

Figure 9.1 *Scenario context: Decreasing dwelling heat-loss factors*

Figures 9.2 and 9.3 illustrate how increased energy efficiency will change heat loads in a typical building now and in a future insulated dwelling that is maintained at low temperatures. Increased energy efficiency will reduce the seasonal variation in heat demand. However, air-conditioning needs will increase if house design is poor. These heat load variations are shown to illustrate how the supply requirements for heating fuels – gas, electricity and CHP heat – depend upon many demand factors. Note that in Figure 9.2 the thermostat setting refers to the temperature below which space heating is activated (there is a different upper limit above which air conditioning is activated). The particular space heat setting of 20°C is one where it is assumed that system control has reduced the average dwelling comfort temperature (this is part of an analysis of the effects of occupant behaviour).

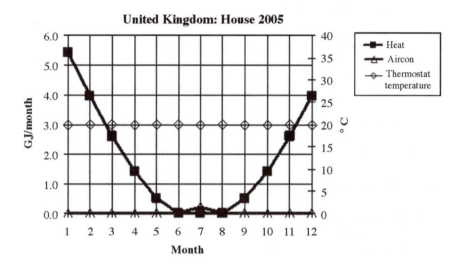

Source: SEEScen model

Figure 9.2 *Scenario context: Current dwelling monthly heat demands*

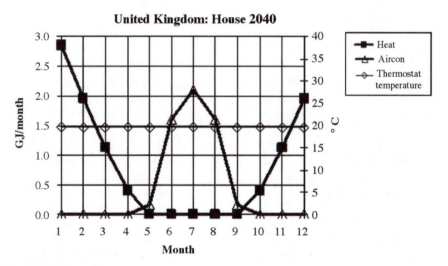

Source: SEEScen model

Figure 9.3 *Scenario context: Insulated dwelling monthly heat demands*

A SUSTAINABLE ELECTRICITY SYSTEM

The bulk of the remainder of this chapter describes the construction of a scenario for a sustainable integrated electricity system for the whole of the UK. The development profile over the years of the system, say 2006 to 2040, is not described here; if the 'end state' system appears feasible based on current technology and incremental extensions, there seems to be no reason to doubt that the system could be constructed over the coming decades.

The components of the system are described: demands, generators, stores and trade links. These are then assembled into an optimized system that enables demands to be met securely at least cost. This involves modelling the hourly variation in demands and supplies, and adjusting the capacities of the different sources and stores until the minimum cost is found.

Spatial issues are not explicitly addressed here. Increasing the geographical range of electricity systems increases the temporal diversity of demand and supply, which reduces the temporal problem of matching renewable energy to demands, but also imposes spatial requirements for transmission and distribution networks. These requirements and their associated costs are not analysed here.

Electricity demands

Future electricity service demands in terms of type, temperature, quantity, time and weather dependency rely on the scenario context. In this scenario, it is assumed that the demands for electricity-specific services that can use no other fuel (such as computers, communications, lighting and some motive power) will generally stabilize because efficiency gains will offset growth factors such as population.

The use of electricity for heating depends upon the relative availability and prices of other fuels that can perform this function. Energy efficiency will reduce the heat demand in dwellings; but gas will eventually have to be replaced, perhaps with electric heat pumps. The balance of these effects on electricity demand requires analysis.

The replacement of liquid fossil fuels is perhaps the most difficult energy supply problem. Electric vehicles are here assumed to make large inroads into the car and light haulage vehicle markets. Vehicles are mostly in use during times of high electricity demand; therefore, their batteries would be predominantly charged at off-peak times, thus reducing the diurnal variation in total electricity demand. The use of electricity for making hydrogen to be used in fuel cells is excluded because this route is currently less efficient than the combination of batteries and electric motors. If hydrogen technologies were to improve, hydrogen fuel cells would replace batteries, where appropriate. This substitution would not change the rest of the electricity system significantly because the electricity requirements for charging batteries or producing hydrogen would be similar in quantity.

Other characteristics of demand can also be important:

- *end-use technologies:* the capacity for control, storage and the ability to be interrupted.

- *multi-fuelled energy services:* electric, solar and gas heating, and electric and liquid-fuelled hybrid cars; fuels can be switched according to which energy supply is available.

Electricity service demands are divided into six categories, each with different weather dependencies, use patterns and service storability. An electricity demand forecast of 282 terawatt hours (TWh) is used, as shown in Table 9.1. This varies from year to year because of weather. The possibility of meeting larger demands is discussed at the end of the chapter.

Table 9.1 *Future annual UK electricity demand characteristics*

Service	Service storable	Weather dependent	Electricity service demand (TWh)	Percentage of demand
Lighting	No	Yes	22	8%
Non-space heating	Yes	Slightly	70	25%
Space heating	Yes	Yes	34	12%
Air conditioning	Yes	Yes	4	2%
Electric vehicle charging	No	Slightly	37	13%
Electricity specific	No	Slightly	114	40%
Total electricity service demand			282	100%

Source: SEEScen model

Energy storage

Energy storage is used to improve the match between demand and supply. From early in the development of electricity systems, energy storage has been used to reduce the peak capacity requirements for transmission, distribution and generation. Since the 1950s, storage has been installed in the electricity system and extensively in consumers' buildings.

System storage, currently in the form of pumped storage, is used to help even out the load on generators and to provide fast response in case a large generator fails. Energy is put into the store by pumping water uphill and is taken out by letting the water back out through hydroelectric turbines. Currently, pumped storage in the UK has a capacity of some 2 gigawatts (GW) electrical power and can store some 10 gigawatt hours (GWh) of energy.

To increase base-load demand and thereby improve the economics of large coal and nuclear generators, off-peak heating was introduced to consumers' premises in the form of hot water tanks and off-peak storage heaters. Currently, some 10 per cent of UK dwellings have electric space heating, and a large fraction have electric heaters (immersion heaters) in hot water tanks, although most water heating uses gas. These existing systems incorporate stores that have a total UK electrical input power of perhaps 40 gigawatts of electricity (GWe) to 60GWe, and a maximum energy storage capacity of some 200GWh.

Further end-user storage can be implemented – for example, in end-use technologies such as building thermal mass, refrigerators or batteries in electric vehicles.

Generation

Renewable electricity sources

Cost and performance figures for renewable technologies have been taken from a number of sources. For most of these, the cost reductions that might occur over the next 20 to 40 years have to be projected, with uncertainties being particularly large for wave and tidal power for which there is little commercial experience, and for photovoltaics where there may be step changes in technological improvement. The total economic energy resource available from renewables depends largely upon the availability and cost of competing fuels – the greater the cost of other fuels, the greater the 'economic' renewable resource. This is particularly so for wind and wave, which have large offshore resources, and solar heating and photovoltaics. There are, however, narrower technical limits to some renewable sources – most notably, hydro and waste biomass.

Table 9.2 summarizes some costs and estimates for the potential of renewable electricity. The table is mainly based on the working paper *Technical and Economic Potential of Renewable Energy Generating Technologies: Potentials and Cost Reductions to 2020*, an input by the UK Cabinet Office's Performance and Innovation Unit (PIU) to the UK government's 2003 Energy Review (PIU, 2002). Of particular note is the projected cost of photovoltaics (PV) mounted on buildings for 2020. The tidal lagoon estimates are quoted in *A Severn Barrage or Tidal Lagoons?* (FoE Cymru, 2004). The potentials and costs have been used as a guide to the inputs of the later modelling, although the costs assumed in the modelling are generally higher than in Table 9.2, which reflects caution in using cost projections. However, it is to be noted that a 95 per cent renewable system would be fully implemented some years after 2020 when costs will have been further lowered because of technology development and mass production.

Biomass is a renewable energy source that can provide heat and electricity at any time because the energy is in a stored form. Table 9.3 shows estimates of the mass and energy content of waste and energy crop biomass. The amount of electricity that may be generated from this in CHP plant has been calculated using an efficiency to electricity of 25 per cent and a capacity factor of 45 per cent.

Combined heat and power (CHP)

CHP converts fuel into heat, which is then used to produce steam and electricity, and the waste heat is used to meet heat demands. In consequence, CHP uses about two-thirds as much fuel as providing electricity and useful heat separately, with CO_2 emissions reduced accordingly. The potential output of CHP depends upon low-temperature heat demands, and the fraction of these demands that might be met with available, cost-effective combustible fuels or other source of high temperature heat.

Table 9.2 *Renewable energy technical and economic potential*

Source	Technology	Cost (UK pence/kWh)	Economic potential at this cost (TWh)	Technical potential (TWh)	Capacity factor (percentage)	Economic potential at this cost (GW)	Technical potential (GW)
Solar	Building photovoltaics	7.0	1	266	14	0.4	216.9
Wind	Offshore	2.8	100	3500	30	42.3	1479.8
Wind	Onshore	3.0	58	317	27	22.1	120.6
Wave		4.0	33	600	40	9.4	171.2
Tidal	Stream	7.0	2	36	40	0.5	10.3
Tidal	Lagoon	2.5	24	24	61	4.5	4.5
Hydro	Small	7.0	2	40	80	0.3	5.7
Biomass	Municipal waste	7.0	7	14	60	1.2	2.6
Biomass	Landfill gas	2.5	7	7	60	1.3	1.3
Biomass	Energy crops	4.0	33		70	5.4	0.0
Total			266	4804		87.4	2012.9

Source: PIU (2002)

The current UK CHP electricity output is about 27TWh from 5.6GWe capacity, running at a capacity factor of 55 per cent. Some estimates, such as *The Government's Strategy for Combined Heat and Power to 2010* (DEFRA, 2004) place the potential to be in the range 50TWh to 100TWh for large-scale and micro-CHP, corresponding to about 20GW capacity. However:

- A major fraction of this CHP would use imported fossil fuels (mostly gas), which will become scarce and expensive and are carbon-emitting.
- This assumes heat loads that may actually be smaller in scenarios with high energy efficiency: CHP potential depends upon the overall scenario context.

Table 9.3 *Biomass potential*

Biomass		Mega-tonnes (Mt)	Gigajoule (GJ)/tonne	Petajoule (PJ)	Eff(e) (percentage)	Terawatt hours of electricity (TWhe)	Capacity factor	Gigawatts (GW)
Waste	Wood waste	4.5	13	59	25	4	45%	1.0
	MSW	8.0	9	72	25	5	45%	1.3
	Straw	3.0	14	42	25	3	45%	0.7
	Sewage	0.4	15	6	25	0	45%	0.1
	Animal waste	3.0	7	21	25	1	45%	0.4
	Total	18.9		200		14		3.5
Energy crops		8.0	14	112	25	8	45%	2.0
	Total	26.9	10	312		22		5.5

Note: MSW = municipal solid waste.
Source: MacLeod et al (2005)

Accordingly, in the scenario it is assumed that CHP potential is ultimately limited to that which may be fuelled with biomass – 5.5GW at a 45 per cent load factor, producing 22TWh of electricity. However, during the period leading to a fully sustainable, renewables-based system, gas and other fuels used for heating and electricity should be used in CHP plant, where possible, because this results in lower fuel use and CO_2 emissions. A scenario in which fossil-fuelled CHP increases and then declines over the next 40 to 50 years may be envisaged. Heat distribution networks developed for CHP would facilitate the rapid and economic introduction of other heat sources, such as electric heat pumps or solar energy, as gas supplies become scarce.

Variations in the electricity and heat outputs of CHP are basically determined by variations in the heat loads they service across the day, week and year. The electricity output of CHP may be manipulated by:

• altering the heat to electricity ratio of the generator (within a range that depends upon the type of technology);
• using heat storage to decouple the CHP heat output from heat demand (this allows CHP to contribute to some electricity demand–supply matching).

Optional generation

Optional generation is electrical generation that can be provided depending upon whether variable generation, storage and trade are insufficient to meet demand. It is possible to avoid any such generation by increasing the capacity of storage and international links. However, further analysis is needed to establish whether this would be economically optimal. In the optimized system, optional generation operates in an annual capacity factor range of 5 to 20 per cent. Existing and new fossil-fuelled generation could be used to meet any deficit of CHP and renewable electricity supply. Currently, there are about 55GW of capacity in major fossil stations, and about 10GW available from private generators as shown in Table 9.4. Some of these could be retained for the long-term future or new flexible plant could be built, depending upon the economics. The coal-fired stations provide strategic security since they can use indigenous reserves.

Table 9.4 *Current UK firm capacity (optional generation)*

Type	Fuel	GW	Future fuel supply
Public (large)	Coal	19	Large domestic coal reserve
	Oil	5	Imported oil held in strategic reserves
	Dual-fired	6	
	Gas	25	Imported gas, some held in UK storage
Private (small)		~10	

Sources: various

Regional and international linkage

Currently, it is assumed that there is sufficient diversity within the continental European system that the UK can import or export, at any time, to the capacity limit of the international UK–France link. Here, this is set at a maximum of 6 GW, compared to the current 2 GW. This is a strong assumption; but the capacity of interconnection across and outside Europe will almost certainly increase.

SYSTEM INTEGRATION AND OPTIMIZATION

Operational issues

The design and operation of electricity systems depend upon the reliability with which demands and supplies can be predicted over different time scales, from minutes to months, and the sophistication of the control of demand and supply technologies. As the UK electricity system becomes increasingly integrated with the European system and systems further afield, the design and operation of these systems will become more complex.

Communications and information processing technologies are already adequate for the precise control of demand technologies, generators and storage. The accurate prediction of demand and supply will become more critical, mainly because of the larger variable renewable component of supply and its consequences for the operational management of demand, storage and trade. However, prediction will improve with the refinement of weather and other data, and of simulation models:

- Weather forecasting will become more precise.
- Energy efficiency and demand management will reduce the less predictable weather-dependent loads, such as space heating and lighting.
- Demand prediction will become more accurate as models improve.
- Predicting outputs from variable sources will become more accurate.

Demand and supply correlation

The planning of electricity supply must include detailed demand analysis because:

- Weather variables are correlated.
- Energy demands vary with time because of social activity and weather patterns.
- Renewable energy supply is weather dependent.

The firm capacity of renewables is the capacity of optional (biomass, fossil or nuclear) sources replaced, such that demands can be met with a specified reliability. The firm capacity of a renewable source depends on the correlated variations in demands and renewable supplies. Importantly, the variations in some demands depend upon the same weather parameters as the outputs from some renewables. To illustrate:

- If solar photovoltaics were to meet space heating demand, its firm capacity would probably be close to 0 per cent; if it were to meet air-conditioning demand, it might be 50 per cent because large air-conditioning loads occur at times of high insolation.
- Air-conditioning load is negatively correlated with wind speed. However, a significant fraction of space heating is positively correlated to wind speed and wind power because wind increases ventilation losses. Assuming that a significant fraction of buildings use electric heat pumps, then the electricity demand from these houses would vary by several gigawatts with wind speed (note that this ignores time lags due to building thermal mass and differences in location of demand and supply).

Load management

Load management is the process of manipulating demand by means of storage and interruption in order to better match supply. A portion of electricity demand may be moved if the net cost of a move is negative, accounting for differences in marginal supply costs, energy losses and other operational costs.

Electricity demand may be disaggregated into segments across sectors and end uses, each segment with a temporal profile and load management characteristics, such as energy storage capacity. Variable electricity supply comprises renewable sources and heat-related generation, each with their own temporal profile. The mismatch between variable sources and demand can be met with a combination of optional thermal generators (characterized by energy costs at full and part load, and for starting-up), traded import or export, and system or end-use storage.

Figures 9.4 and 9.5 demonstrate the role that load management might play in a putative future system (different from the 95 per cent renewable system described later in this chapter), integrating variable renewables and CHP within electricity supply on a winter and summer's day, shown in the graphs as starting at 0 hours. Heat and electricity storage (hot water tanks, storage heaters and vehicle batteries) can be used to store renewable energy when it is available so that the energy can later be used when needed. Other demands, such as refrigerators, can be manipulated or interrupted. In this example of load management, mainly heat demands are managed with storage.

Figures 9.4 and 9.5 show how, by moving demands with storage, the system demand profile can be matched to variable supply from CHP and uncontrollable renewables. The residual demand to be met by optional generators (conventional nuclear and fossil) is then flat. This means that these plants do not have to 'load follow', which wastes energy, and that the required installed capacity of such plant is reduced.

The system graph (Figures 9.4 and 9.5, top left) summarizes:

- system demand (end-use demand plus transmission losses);
- variable renewable and CHP supply (here, called 'essential');

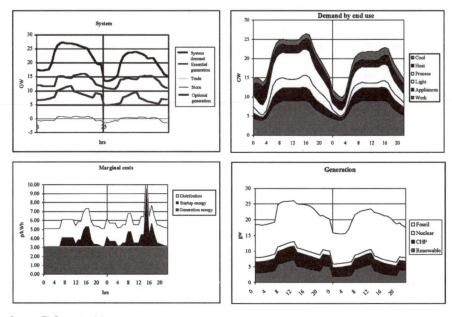

Source: EleServe model

Figure 9.4 *Matching without load management: A winter and summer's day*

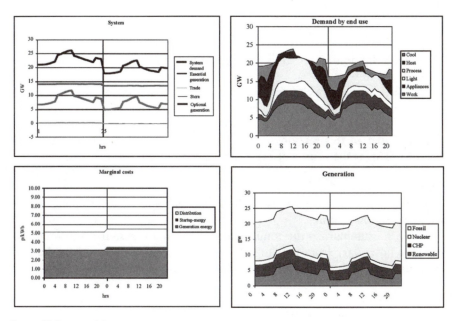

Source: EleServe model

Figure 9.5 *Matching with load management: A winter and summer's day*

- trade;
- system storage (pumped storage);
- optional supply required to meet difference between demand plus storage and variable supply.

The demand graph (Figures 9.4 and 9.5, top right) shows demand for different end uses: cooling, heating, processes, lighting, appliances, and work or motive power. The marginal costs graph ((Figures 9.4 and 9.5, bottom left) shows the energy costs of generation, the costs of starting up plant and the distribution costs (currently, a simple constant).

The generation graph ((Figures 9.4 and 9.5, bottom right) shows the output from each generation source in sequence from bottom to top: renewable sources, CHP and optional generation ordered by increasing steady-state marginal cost (excluding start-up costs).

The load management simulation in Figure 9.5 demonstrates how variable electricity supplies constituting about 50 per cent of peak demand, using heat storage alone, can be absorbed so that the net demand met by optional generation is levelled. This indicates that large fractions of variable electricity supply can be absorbed into the electricity system without special measures other than the control of heat stores. Further investigation of other system configurations and renewable generation is required to establish the exact potential of load management.

An optimized system: Summary

A model called Energy Space Time (EST) is used to:

- simulate the hourly demands, renewable and optional generation, storage and trade flows;
- find the least-cost mix of generators, stores and trade capacities. The annuitized costs of capital are calculated using a 5 per cent discount rate.

The remainder of this section summarizes the optimized system as modelled with EST. This is the source of tables and graphs in this section.

Technical

Figure 9.6 depicts the capacities of the generators and trade link, and of the electricity and heat stores.

Table 9.5 summarizes the annual energy flows of the system for a simulated year. The flows vary from year to year because of fluctuations in demand and renewable output due to variations in the weather and renewable resources.

Economics

Table 9.6 shows the total annual cost of the system and the average unit price of electricity. Both of these will vary from year to year because of weather-induced changes in demand and in supply, particularly in trade and optional generation. The annual cost of the renewables does not change significantly

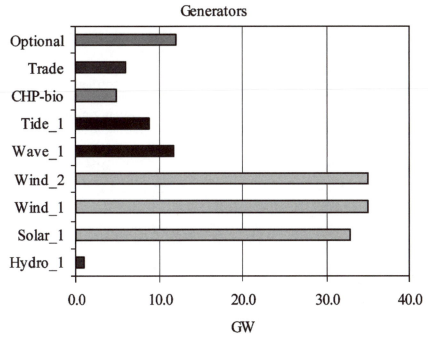

Note: Key on page 174.
Source: EST

Figure 9.6 *Generator capacities*

Table 9.5 *Annual energy: Technical summary*

		TWh	
Demand		282.2	
	Transmission losses	16.9	
	Supply requirement	299.1	
Supply	Renewable	292.2	98%
	Spilled	−10.0	−3%
	CHP-bio	19.2	6%
	Optional	5.2	2%
	Storage	2.2	1%
	Country supply	308.8	103%
	Country surplus	9.7	
	Trade	−8.8	
	Country supply	300.0	

Notes: 'Renewable' refers to electricity-only renewable systems. 'CHP-bio' refers to biomass-fuelled CHP. 'Spilled' is the electricity spilled because it is generated by renewables but cannot be absorbed by demand, storage or export.
Source: EST

except for expenditure on maintenance that is related to energy output for that year. Negative energy costs arise because of export. On average, the trade balance for the optimized system is near to zero.

Table 9.6 *Annual costs: Economic summary*

Annual cost	£UK
Capital	16.7
Energy	−0.7
Store	0.3
Total	16.2
Average	5.4 UK pence/kWh

Source: EST

The pie chart in Figure 9.7 shows the distribution of annualized expenditure summarized in Table 9.6.

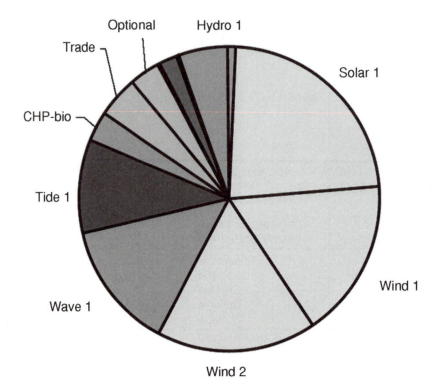

Note: Key on page 174.
Source: EST

Figure 9.7 *Breakdown of annualized component costs*

Table 9.7 Optimized system details

		Demands						Supply — Renewables									Storage — Electricity			Storage — Heat		
		Light	Heat	Space heat	Air con	EV charge	Ele spec	Hydro_1	Solar_1	Wind_1	Wind_2	Wave_1	Tide_1	CHP-bio	Trade	Optional	Stel_In	Stel_Sto	Stel_Out	Sthe_In	Sthe_Sto	Sthe_Out
Capacity Current	GW	2.5	8.0	4.0	0.5	4.0	13.0	1.0	32.9	35.0	35.0	11.7	8.8	5.0	6.0	12.0	4.6	10.0	6.3	24.9	300.0	0.0
Maximum	GW							1.0	55	35	35	12	10	5	6	50	100	400	100	100	999	100
Minimum	GW							0.6	0	0	0	0	0	2	2	0	2	10	2	0	10	10
Efficiency								86%	25%	25%	25%	60%	60%		92%		88%	77%	88%	99%	97%	98%
Energy	TWh	22	69	35	5	37	114	7.5	41	93	84	35	32	19	-9	5						
Capacity factor								86%	14%	30%	27%	34%	42%	44%		5%	-5%		3%	-6%		5%
Unit capital cost	£/kw							2500	2000	1000	1000	2500	3000	1000	1500	200	100	400	100	10	50	5
Operating life	Yrs							100	30	20	20	20	25	25	50	35	20	20	20	30	30	30
Discounted life								19.8	15.4	12.5	12.5	12.5	14.1	14.1	18.3	16.4	12.5	12.5	12.5	15.4	15.4	15.4
Capital total	G£							2.5	65.8	35.0	35.0	29.2	26.5	5.0	9.0	2.9	0.5	4.0	0.6	0.2	15.0	0.0
Capital annuitized	G£							0.1	4.3	2.8	2.8	2.3	1.9	0.4	0.5	0.2	0.0	0.3	0.1	0.0	1.0	0.0
O&M cost	£/kW/a							25.0	20.0	20.0	20.0	50.0	30.0	20.0	30.0	4.0	2.0	8.0	2.0	0.1	0.5	0.0
O&M cost	G£							0.0	0.7	0.7	0.7	0.6	0.3	0.1	0.2	0.0	0.0	0.1	0.0	0.0	0.2	0.0
Energy cost (O&M, fuel)	p/kWh							0.1	0.1	0.1	0.1	0.1	0.1	1.0	10.4	6.9	0.0	0.0	0.0	0.0	0.0	0.0
Energy cost	G£							0.0	0.0	0.1	0.1	0.0	0.0	0.2	-0.9	0.4	0.0	0.0	0.0	0.0	0.0	0.0
Total cost	G£							0.2	5.0	3.6	3.6	3.0	2.2	0.6	-0.2	0.6	0.0	0.4	0.1	0.0	1.1	0.0
Unit cost	p/kWh							2.1	12.1	3.9	4.3	8.4	6.8	3.3	-2.7	11.2	0.0	0.0	0.0	0.0	0.0	0.0

Note: G£ = £billion UK; p/kWh = UK pence per kWh.
Source: EST

Optimized system: Demand and technology details

Table 9.7 shows further details of the optimized system. Shaded cells contain assumptions; those with bold type are the values changed by the optimizer. The rows labelled minimum and maximum show the allowed range of values, and the row labelled current is the optimized value between those limits. Of note are the energy generated, the capacity factors, and the unit cost of electricity generation shown in the last row. The negative numbers for trade arise because, for this particular year, electricity is exported.

Key to Figures 9.6–9.12 and Table 9.7

Demand	Light	Lighting
	Heat	Water and other heating
	Space heat	Space heating
	Air con	Air conditioning
	EV charge	Electric vehicle charging
	Ele spec	Electricity specific
Supply	Hydro_1	Hydro 1 generation
	Solar_1	Solar PV 1 generation
	Wind_1	Wind 1 generation
	Wind_2	Wind 2 generation
	Wave_1	Wave 1 generation
	Tide_1	Tidal 1 generation
	CHP-bio	Biomass CHP
Summary	Trade	Trade with France
	Optional	Fossil generation
	Sup_Req	Required supply
	Sup_Var	Variable generation
	Sup_Tot	Total generation
	Tr_Loss	Transmission loss
Storage	StEl_In	Electricity storage input power
	StEl_Sto	Electricity storage capacity
	StEl_Out	Electricity storage output power
	StHe_In	Heat storage input power
	StHe_Sto	Heat storage capacity
	StHe_Out	Electricity storage output power

Optimized system: Hourly performance

This sub-section illustrates how the optimized system performs hourly for individual sample days, and throughout the year. Figure 9.8 shows a winter's day in which the variable supply of electricity from renewables and CHP is greater than demand during the day. Surplus variable generation is exported and placed into energy stores.

Demands and supply

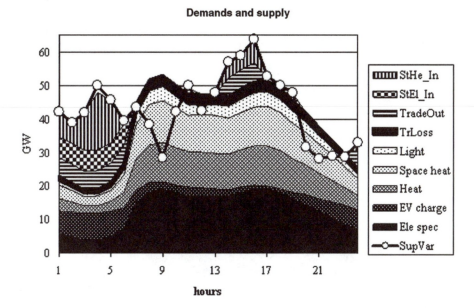

Supplies and demand

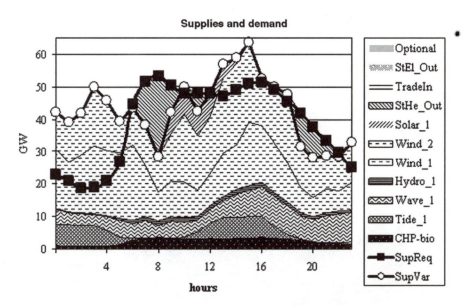

Note: Key on p 174.
Source: EST

Figure 9.8 *Sample winter's day: Variable supply excess*

Figure 9.9 shows a winter's day when demand is greater than variable supply during the day. The deficit is met with import, energy from stores and optional generation.

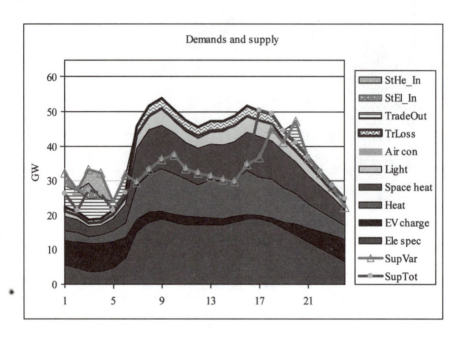

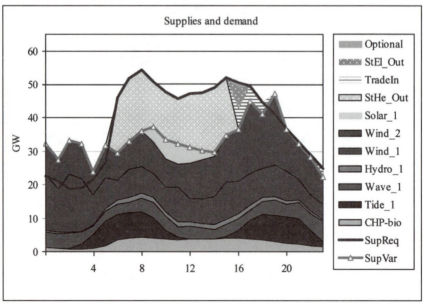

Note: Key on p 174.
Source: EST

Figure 9.9 *Sample winter's day: Variable supply deficit*

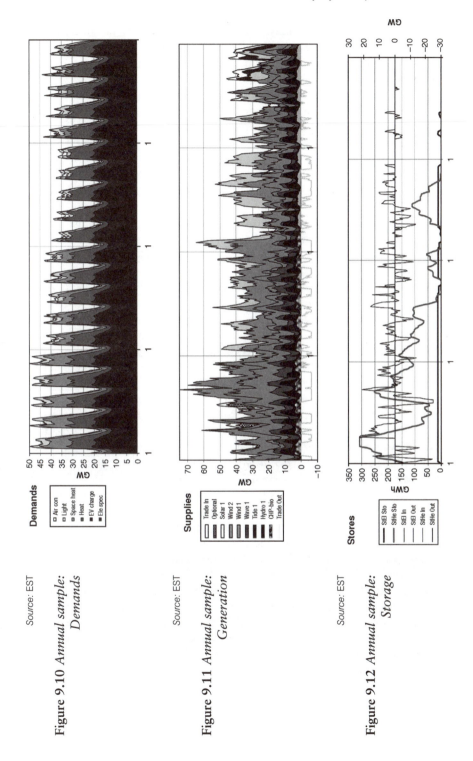

Figure 9.10 *Annual sample: Demands*

Figure 9.11 *Annual sample: Generation*

Figure 9.12 *Annual sample: Storage*

Figures 9.10, 9.11 and 9.12 show the electricity system performance for four sample months (January, April, July and October), each with five sample days. Figure 9.10 shows how the assumed demands vary due to socio-economic activity and weather patterns.

Figure 9.11 shows the generation from renewables, CHP and optional sources: trade imports are shown as positive, exports as negative.

Figure 9.12 depicts the energy stored (thick lines) and inputs and outputs from the stores.

CONCLUSIONS[1]

This chapter has summarized some of the principal features of a sustainable electricity service system. It has shown how indigenous renewable energy sources can provide up to 95 per cent of electricity supply securely if storage, trade and optional generation are also deployed. It further demonstrates that the unit cost of electricity (averaging about 5.5 UK pence per kilowatt hour) may not be excessive when compared to future fossil or nuclear generation costs. The chapter emphasizes that the cost estimates for renewables in 10 or 50 years time are inevitably speculative, as they are for fossil and nuclear generation. However, there is more certainty about renewable energy costs because they are not dependent upon finite fuel prices, which will inexorably increase. The future unit cost of electricity will probably be higher than today in any scenario because of capital cost and fuel price increases, either in a high renewable future or one with large fractions of fossil and nuclear generation. The scenario is more secure than high fossil/fissile scenarios because it is almost immune to the unpredictable future prices and availabilities of finite fuels, and it incorporates a mix of low-risk, reversible technologies.

It is not claimed that this system is necessarily the best since it does not include all the possible options in terms of technologies or operational strategy. The optimal system depends upon the many assumptions about future demands, and the performance and costs of generation, storage and transmission technologies. Changes in these assumptions will lead to different solutions. However, some of the costs and technicalities of a working system have been demonstrated: the challenge is to find better solutions.

Energy security can be defined as the maintenance of safe and economic energy services for social well-being and economic development, without excessive environmental degradation. Demand management and energy efficiency are the fundamental options to improve security. Most forms of energy supply are associated with some combination of technical, environmental or economic insecurity:

- Renewable sources are, to a degree, variable and/or unpredictable. However, most renewable technologies are dispersed, mass-produced, reversible (they can be removed without trace) and present no large-scale risks.
- Fossil fuels produce greenhouse gases, are finite and will be increasingly imported. UK coal reserves are large; but coal has a high carbon content.

Imported fuels suffer price volatility.
- Nuclear fuels are finite, and nuclear technologies are, effectively, irreversible and present potentially large risks.

Further analysis

This chapter presents the results of work in progress. There are many aspects that warrant further investigation in order to test the robustness of the system, and to seek different and better solutions.

Increased demand

The demand for electricity is the fundamental driver. Careful analysis of this is required, especially in the context of overall energy, in which electricity may substitute for gas in the stationary sector and for liquid fuels in the transport sector. The correlations, positive or negative, between demand and renewables have significant impacts on costs.

If demand is increased, then so must supply. The PIU (2002) survey of renewable energy sources shows that to increase renewable output to 400TWh to 500TWh per annum, or more over the next 40 to 50 years, may not be an unreasonable target. The offshore wind and building-integrated photovoltaic resources, using proven technologies, are in excess of 1000TWh per year. As renewable supply increases, so the average unit cost of supply rises; but modelling indicates that the rate of cost increase is not very steep.

Different renewable fractions

Beyond the scenario presented here, systems with 100 per cent renewable energy have been modelled and shown to be feasible. With such a penetration, the detailed analysis of demand–renewable correlations, renewable siting and technologies, and storage and trade becomes more critical.

International electricity context

An extensive continental grid already exists, and increases in the capacity of connection between the UK system and the continental grid enhance the benefits of diversity at the cost of transmission. Some implications of a larger grid are discussed elsewhere in this book. The advantages of extending the system include more demand and renewable supply diversity because of different weather and demand patterns (the latter includes the effect of time zones) in other countries. Some European countries have a large hydro component, which is, to a degree, an optional renewable source and may be used for some matching of generation to demand.

Economics

If the performance and relative costs of the technologies changed, then so would the configuration of the optimum system. Assumptions about storage and photovoltaic technologies are perhaps the most critical.

Arguably, photovoltaic generation has the least environmental impact of the renewable sources. In addition, most PV would be sited near demand, on buildings, where maintenance and transmission needs and costs would be less than for the remote sources. An interesting question, then, is to what extent a reduction in the relative cost of PV would increase its contribution in an optimized system.

Modelling of the following aspects could be refined:

- demand, in terms of quantity, use patterns, weather dependency, control and correlation with renewable supply;
- technologies and their controls (i.e. load management, storage and renewable energy);
- the spatial aspects of the system and transmission requirements.

Optimization could include:

- demand management decision variables, such as the ability to be interrupted and efficiency costs;
- control strategy parameters for operating demand management, storable renewables (hydro and tidal), stores, CHP and trade.

NOTE

1   More details and context of the work presented in this chapter may be found at: www.bartlett.ucl.ac.uk/markbarrett/Index.html.

REFERENCES

DEFRA (2004) *The Government's Strategy for Combined Heat and Power to 2010*, Crown Copyright, www.defra.gov.uk/environment/energy/chp/index.htm

FoE Cymru (2004) *A Severn Barrage or Tidal Lagoons?*, www.foe.co.uk/resource/briefings/severn_barrage_lagoons.pdf

MacLeod, N., Clayton, D., Cowburn, R., Roberts, J., Gill, B. and Hartley, N. (2005) *Biomass Task Force Report to UK Government*, www.defra.gov.uk/farm/crops/industrial/energy/biomass-taskforce/pdf/btf-finalreport.pdf

PIU (Performance and Innovation Unit) (2002) *The Energy Review: A Performance and Innovation Unit Report Technical and Economic Potential of Renewable Energy Generating Technologies: Potentials and Cost Reductions to 2020*, UK Cabinet Office (now Prime Minister's Strategy Unit), London

10

Reliable Power, Wind Variability and Offshore Grids in Europe

Brian Hurley, Paul Hughes and Gregor Giebel

WHERE ARE THE WIND RESOURCES?

This chapter takes as a starting point the recent study *Sea Wind Europe* (Greenpeace, 2004), which identifies the magnitude of the offshore resource by country, the annual yield in gigawatt hours (GWh), the capacity in gigawatts (GW), the area occupied in square kilometres (km²), and the percentage of available area for the periods of 2003–2010, 2011–2015 and 2016–2020.

A recent report by the European Wind Energy Association (EWEA, 2004) provides further material and, with earlier studies, gives estimates of the land-based potential for wind energy development. The total on-land technical potential yield for the 15 European Union member states (EU-15), plus Norway, is given as 649 terawatt hours per year (TWh yr⁻¹).

For offshore wind, a 1995 study by Garrad et al (1996) gives the technical potential as 2463.7TWh yr⁻¹ within some tens of kilometres off the coasts of Europe.

These totals are very significant in comparison with the current generation within the EU-15 of 2572TWh yr⁻¹ (2000 figures). A more detailed measure of the scope offshore can be gleaned from an examination of Table 10.1 from the Greenpeace (2004) report.

European winds are driven by the westerly atmospheric circulation that is characteristic of middle latitudes. Frontal systems and depressions are a feature of this westerly circulation. A longer-term influence is the North Atlantic oscillation, which has a large climatic influence on the North Atlantic Ocean and the surrounding landmasses. This has an effect on the tracking of weather systems, moving them from a north-easterly track to a more easterly track across the landmass of Europe (see Figure 10.1). The Mediterranean region is influenced by a series of more localized effects (see Figure 10.2).

Table 10.1 *Possible offshore development within Europe (2020)*

	Annual energy yield (AEY) (GWh)	Capacity (GW)	Area occupied (km²)	Percentage of available area
Belgium	23,077	6.67	834	37.37
Denmark	95,126	27.79	3474	3.98
Finland	18,366	13.40	1675	2.66
France	106,065	32.78	4097	6.27
Germany	40,766	11.54	1443	5.47
Greece	2755	3.30	413	1.91
Ireland	56,935	15.34	1917	3.19
Italy	26,014	16.98	2122	4.36
Netherlands	24,046	6.56	820	1.62
Portugal	39,188	12.74	1592	16.07
Spain	77,831	25.52	3190	9.57
Sweden	47,161	17.26	2157	1.99
UK	163,566	46.75	5844	1.97
Total	720,896	236.62	29,578	3.38

Source: Greenpeace (2004)

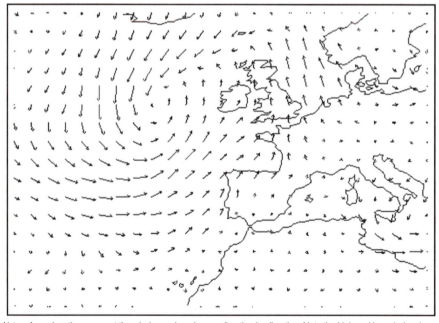

Notes: Arrow lengths represent the wind speed, and arrow direction its direction. Note the high and low wind regions.
Source: www.cdc.noaa.gov/cdc/reanalysis/reanalysis.shtml

Figure 10.1 *Typical wind patterns over North-West Europe*

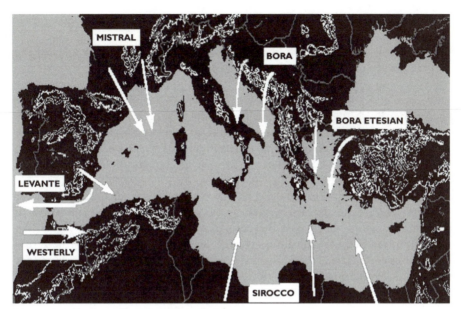

Source: www.nrlmry.navy.mil/~medex/medmap.html

Figure 10.2 *Wind patterns over the Mediterranean region*

WHAT HAPPENS WHEN THE WIND DOES NOT BLOW?

It is a rare event when the wind does not blow in a region, and an even rarer event when a larger region is considered. Taking Europe as a whole, it has been found that averaging 51 one-year onshore time series on an hour-by-hour basis leads to a minimum generation of 1.5 per cent of the installed capacity at all times. But how will a region cope with the rare event of a lack of wind or with low wind? One has to consider what occurs as more wind is integrated. Several parallel developments are likely to transpire as part of policies to further increase the penetration of wind:

- A different mix of conventional plant will be the context as new conventional plant is added – for example, more open-cycle gas plant or combined-cycle plant designed for flexibility would become available.
- Older conventional plant, such as coal, oil and gas plants plants being retired, can be refurbished so that they may be used as a cheap form of 'storage', either being called on occasionally to run from conventional fuel or new liquefied natural gas supplies, or (with suitable modification) to use new renewable fuels, such as solid or liquid biomass. In the longer-term, hydrogen and other suitable energy storage media will also be available.
- Additional interconnection will become available as inter-country and inter-regional flows increase to allow greater flexibility in meeting maximum demand in specific regions.

- There will be specific interconnection of regions to capture the geographical dispersion effect for wind.
- There will also be increased information technology enhancement of the grid (i.e. an emergence of the 'intelligent' grid).

Low correlations between wind power production at individual sites that are geographically distinct will smooth the production profile since the probability of some wind generation at any time is higher when the production profile does not rise and fall in unison at different sites. Typical weather patterns in Europe are only about 1500km in extent. Essentially, this implies that the wind always blows somewhere.

The cross-correlation between any 2 of 51 stations distributed all over Europe was investigated. From these (onshore) stations, one year's worth of three-hourly wind speed and direction data was available (Giebel, 2001), typically measured at 10m above ground level. A small value for the cross-correlation coefficient means that the single time series adds up to a smoother time series, while time series with high cross-correlation coefficients just add their variability.

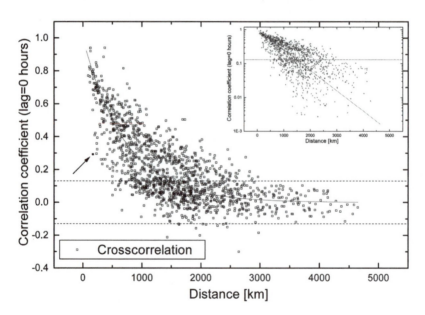

Notes: The dashed lines at ±0.13 are only used to guide the eye. The arrow points to the pair of stations on Sardinia, Italy. In the inset, the same plot is scaled logarithmically. The solid line is an exponential fit exp (–Distance/D), with D being 723km.
Source: Giebel (2001)

Figure 10.3 *Correlation coefficient for every pair of stations at lag = 0 hours*

Figure 10.3 shows the correlations for all pairs of stations, together with their respective distances. While short distances give the highest correlations, a short distance does not necessarily mean that the time series are correlated. Local effects can actually lead to a significant decoupling of the time series (Joensen et al, 1999). The best example for this is the point represented by the pair Alghero–Cagliari in Italy, where the cross-correlation for a distance of 170km drops to 0.29. These stations are in the north-west and south of the island of Sardinia in the Mediterranean and, hence, have rather different microclimates. The low wind speeds at both stations (2.9m s^{-1} and 3.9m s^{-1}) point to local influences dominating the wind.

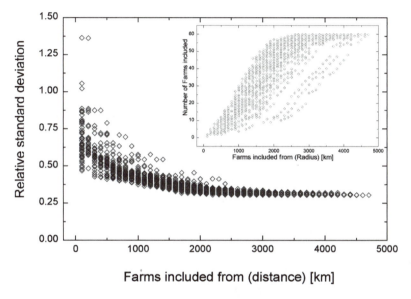

Notes: Relative refers to the standard deviation divided by the mean of the time series. In the inset are the numbers of farms included for a given radius.
Source: Giebel (2001)

Figure 10.4 *Relative standard deviation of the time series resulting from combining all available stations ('farms') within a circle of radius R around any one station*

For longer distances, the result is as expected: the correlation is very small. Hence, spreading out the wind power generators should lead to a reduced variability of the resource since the standard deviation of the sum of N time series is given as:

$$\sigma_{sum^2} = \frac{1}{N^2} \sum_i \sum_j \sigma_i \, \sigma_j \, corr_{ij} \qquad [1]$$

with σ_i and σ_j being the respective standard deviations of the individual time series.

Note that the correlation function of two time series p_t and q_t is as follows:

$$a_k = \frac{1}{N} \sum_{t=1}^{N-k} \hat{p}_t \, \hat{q}_{t+k} \; \Big/ \; \sigma_p \, \sigma_q \qquad\qquad [2]$$

with:

$$\hat{p}_t = p_t - \mu_p \text{ and } \hat{q}_t = q_t - \mu_q. \qquad\qquad [3]$$

The sum μ_p/q is the mean of the corresponding time series; s_p/q is their standard deviation. The time lag between the two series is represented by k. For the autocorrelation function, set $q_t = p_t$.

In Equation 2, a_k is the cross-correlation coefficient. A value of 1 means that the time series are completely correlated, while a value of 0 means that the data are completely uncorrelated.

Averaging these 51 one-year time series on an hour-by-hour basis leads to a minimum generation of 1.5 per cent of the installed capacity at all times (Giebel, 2001). Another study uses the same data, but a different procedure, and comes out with 1 per cent of cases when there is no wind power at all in the European grid (Landberg, 1997). Both these studies refer to onshore only. The wind patterns offshore are less variable than onshore. Our conservative conclusion is that the occasions when the wind does not blow somewhere onshore and offshore are very rare.

A larger catchment area also leads to slower variations in output since the speed of variations is 'washed out' due to the higher frequencies in the wind speeds not being correlated. This is reported, for example, by the Institut für Solare Energieversorgungstechnik (ISET) Renewable Energy Information System on Internet (REISI) (see http://reisi.iset.uni-kassel.de/). Their data show that the maximum changes in electricity production in Germany within one hour in 2002 were, respectively, +20 and –24 per cent of the installed capacity. When aggregating smaller regions, the changes are larger.

Giebel and colleagues show a time series fabricated from reanalysis data (Kalney et al, 1996; Giebel, 2001), considered to be representative of the European average potential wind power generation between 1965 and 1998, as an average over 60 well-distributed sites. A typical feature of the wind energy production is that in summer, wind energy production is much lower than in winter. The different yearly time series are rolled out in Figure 10.5. Every point shown here is one realization of a wind energy production, as averaged over all Europe, within a period of 34 years. More interesting than these points is the empty area surrounding them. In this empty area, in no case during the 34 years analysed did production occur at this level. This includes all the area above 90 per cent generation and above 70 per cent during the summer months, but also the area below 10 per cent for the winter months, where only a very few cases of low wind are seen.

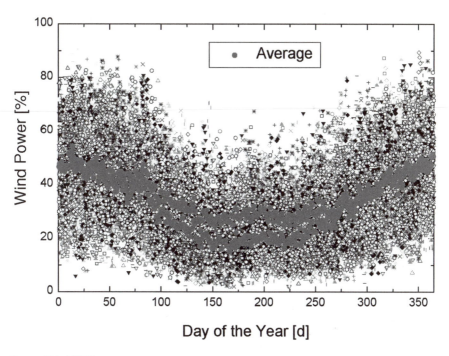

Source: Giebel (2001)

Figure 10.5 *European generation according to reanalysis data: Every single time series corresponds to one year, while the average is the average at every time step of the 34 years*

THE RELIABILITY OF DISPERSED OFFSHORE WIND POWER

In work done within Airtricity, the potential benefits of dispersion and inter-connection of geographically dispersed offshore wind capacity were examined using a likely development strategy for offshore wind. Wind data for seven likely offshore wind farm development locations were extracted from a global weather database for 2003. Coordinates for each location are given in Table 10.2. Wind speeds at 10m were extrapolated to hub height and converted to electrical power via a turbine power curve.

Locations of potential offshore wind development and their coordinates

The resulting series consisted of gross energy production figures at six-hourly intervals. Energy losses were not accounted for in this series (a blanket fixed-

Table 10.2 *Coordinates for potential offshore wind plants*

Location	Latitude (°)	Longitude (°)
Thames Estuary	52.4	1.9
Baltic Sea	56.2	18.8
Celtic Sea	48.6	−9.4
Mediterranean (Marseille)	42.9	3.8
Orkney	60.0	−3.8
Black Sea West	42.9	28.1
Trafalgar	35.2	−7.5
Irish Sea	52.4	−5.6

Source: Hughes and Hurley (2005a, b)

percentage loss would produce a time series that never exceeds the threshold determined by the loss percentage).

The distribution of load factor throughout the year was examined for the seven scenarios. The first scenario placed all capacity in the Thames Estuary. Each successive scenario added an equivalent amount of capacity to the new area until all seven locations had an equivalent capacity. This occurred in the following order: the Thames Estuary, the Baltic Sea, Orkney, the Celtic Sea, Trafalgar, the Mediterranean (Marseilles) and the Irish Sea.

Results

With all wind capacity installed at one location, the frequency of no wind production is around 13 per cent. Periods of full load are also quite frequent, occurring approximately 30 per cent of the time. The distribution of annual load factors has two peaks: one at full load and the other at zero load. This is to be expected, considering the shape of a turbine power curve. This pattern is reflected at all locations in the 'no dispersion' scenario.

As capacity is added successively to each location, the probability of no wind production falls to zero. The distribution of load factors takes on a more Gaussian shape, with just one peak around 55 per cent load factor. The majority of production is clustered around the median value, with two-thirds of all load factors between 30 to 70 per cent of total capacity.

Variability
As dispersion increases, the probability of large changes in power from one period to the next falls to zero. This contrasts with the single location case, where changes of up to 100 per cent of installed capacity can occur. With Europe-wide geographic dispersion of wind capacity across at least six locations, the majority of changes in power are less than 10 per cent of installed capacity.

Table 10.3 *Load factor distribution*

Scenario	Load factor (upper bound) (percentage)									
	10%	20%	30%	40%	50%	60%	70%	80%	90%	100%
1	24.9	10.2	6.7	4.5	5.0	4.7	3.4	4.5	5.2	31.0
2	11.7	9.2	8.3	6.6	14.2	14.3	7.9	6.6	5.5	15.6
3	4.2	6.5	5.2	14.9	13.5	11.4	16.0	9.9	7.5	10.9
4	1.8	4.1	7.6	13.8	12.7	15.5	12.9	15.0	8.4	8.2
5	1.3	3.6	10.0	12.9	18.7	16.0	18.3	11.9	5.9	1.4
6	1.3	5.4	9.2	14.9	16.9	18.6	16.6	11.2	4.9	0.9
7	1.2	5.2	9.3	14.0	17.0	16.7	16.5	12.7	6.3	1.0

data

Scenario	Portfolio
1	Thames Estuary
2	Thames Estuary + Baltic Sea
3	Thames Estuary + Baltic Sea + Orkney
4	Thames Estuary + Baltic Sea + Orkney + Celtic Sea
5	Thames Estuary + Baltic Sea + Orkney + Celtic Sea + Trafalgar
6	Thames Estuary + Baltic Sea + Orkney + Celtic Sea + Trafalgar + Mediterranean
7	Thames Estuary + Baltic Sea + Orkney + Celtic Sea + Trafalgar + Mediterranean + Irish Sea

Source: Hughes and Hurley (2005a, b)

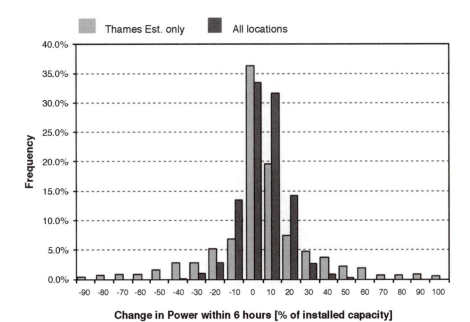

Source: Hughes and Hurley (2005a, b)

Figure 10.6 *The changes in six-hourly output from one wind farm and a distribution of six wind farms*

EFFECT OF WIND FARM POWER OUTPUT FORECASTING

Short-term forecasting of wind farm output has been worked on for more than 15 years within the wind industry (see Chapter 5 and Giebel et al, 2003). It has a function in contributing to making wind more predictable. This allows for the conventional plants to plan ahead to adjust their output appropriately. During earlier years, it was largely of interest to grid operators. More recently, the owners of wind farms have also been taking an interest for two reasons. First, in some regions they are obliged to provide forecasts of output by the grid operator. Second, in some markets there is recognition that the power produced may be of more value if an accurate forecast were available. The period of 1 hour ahead to 72 hours ahead is what is normally technically feasible. For an example forecast, see Figure 10.7.

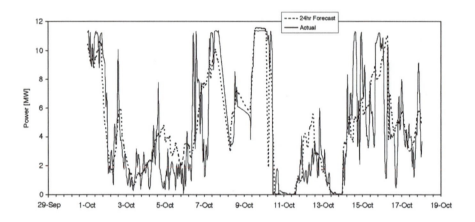

Source: Hurley and Dodhia (2006)

Figure 10.7 *A typical 24-hour (ahead) wind power forecast and the corresponding actual generation*

DELIVERED COST ESTIMATES FOR NEW GRID AND WIND FARMS

For onshore wind farms, the range of costs is taken from the European Wind Energy Association (EWEA, 2004, Figure 2.3, p99). These costs range from €900 to €1150 per kilowatt installed, including grid connection. For the purposes of this analysis, the figure of €1000 is used. For offshore, a cost of €1500 per kilowatt installed is used. In the light of recent reports, this offshore cost is low.

The cost of transmission has been estimated assuming the use of direct current (DC) technology. For power levels of 500MW+ and distances of 100km+, this is the only viable technology. Due to the absence of a synchronous source, it has been assumed that it would be necessary to use voltage source converters at the wind farm end. Although, to date, this technology has not been used at these power levels, there are no obvious technical reasons why this could not be done (PB Power, 2002). At the load end, conventional converter technology would be employed with transmission voltages in the region of ±450 kilovolts (kV).

Costs have been estimated using averages of those quoted by several internal and external sources, including PB Power (2002). Capital costs and losses split between fixed costs (per megawatt) and variable costs (per megawatt kilometre) have been assumed as in Table 10.4.

Table 10.4 *Capital costs*

	Capital costs	
	DC	DC
	Fixed cost	Variable cost
Cost (€ million/MW)	0.378	
Offshore cost (€/MWkm)		630
Onshore cost (€/MWkm)		200
	Losses	
	DC	DC
	Fixed cost	Variable cost
Percentage	2%	
Percentage/100km		0.33%

Source: Hughes and Hurley (2005b)

It has been assumed that annualized costs are equivalent to 10 per cent of capital costs. This results in transmission costs, excluding losses, ranging from under €0.02/kWh to around €0.06/kWh as distances range from 500km to 3000km, depending upon the offshore–onshore balance. This compares with average costs paid for grid capacity of €0.006/kWh from England to France and €0.019/kWh from Germany to The Netherlands in 2003.

While there is likely to be a charge for connecting to the alternating current (AC) grid at the load end, this may be a negative charge if the wind farm output has, as is assumed above, been transported to a major load centre. Therefore, a zero charge has been assumed here as a conservative assumption.

Table 10.5 *Transmission cases and costs*

Transmission cases (1000 MW)

	Distance onshore (km)	Offshore (km)	Total	Total capital costs (€ million)
Scotland to Dublin	175	55	230	448
Dublin to French Alps	1355	45	1400	678
North-West Africa to mid Germany	1500	1	1501	679
Northern Norway to mid Germany	500	1500	2000	1424
Northern Russia to mid Germany	2000	0	2000	778
Greater Gabbard to Köln	290	160	450	537
Baltic to mid Germany	200	200	400	544
Greater Gabbard to The Netherlands	50	200	250	514
South Irish Sea to Köln	800	240	1040	689
North Sea north to mid Germany	300	500	800	753
North Sea south to mid Germany	350	250	600	606
Irish Sea to France	1	500	501	693
Celtic Sea to mid Spain	300	1000	1300	1068

Source: Hughes and Hurley (2005b)

To calculate the cost of transmission of a quantity of electrical energy (in megawatt hours) from a wind farm to a load centre, the utilization factor of the wind farm was set equal to the capacity factor of the wind farm. This is a worst case assumption as the likely utilization factor of parts, if not all, of the transmission line could be much higher since other power could also be transmitted. The capital cost of production per megawatt hour at the wind farm site is calculated using a capacity factor derived from the estimated wind speed at the site (Dowling et al, 2004).

Table 10.6 *Total capital costs at load centres per megawatt hour*

Case	Wind speed (m s⁻¹)	Capital cost generation (per MWh)	Capital cost transmission (per MWh)	Total capital cost at load centres (per MWh)
High wind area (onshore)**	7.07m s⁻¹ at 60m*	379	0	379
Medium wind area (onshore)**	6.45m s⁻¹ at 60m*	450	0	450
Low wind area (onshore)**	5.53m s⁻¹ at 60m*	633	0	633
South Irish Sea to French coast	8.5m s⁻¹ at 100m	512	237	749
South Irish Sea to French coast	10m s⁻¹ at 100m	414	192	606
North Sea South to Germany	8.5m s⁻¹ at 100m	510	206	716
North Sea South to Germany	10m s⁻¹ at 100m	413	167	580
North Sea north to mid Germany	>10m s⁻¹ at 100m	<413	<167	<580
Baltic to mid Germany	8.5m s⁻¹ at 100m	510	185	695
Baltic to mid Germany	10m s⁻¹ at 100m	419	210	629
Thames Estuary	~9m s⁻¹ at 100m	467	160	627

Notes: * Adjusted from 50m with log law r = 0.3.

     ** Mean annual wind speed at 10m; the rest of wind speeds at hub height.

Source: Hughes and Hurley (2005b)

RESULTS FOR EUROPE

What emerges from an examination of Table 10.6, through ranking the cases on the basis of the delivered capital cost of electricity per megawatt hour, is that when medium to high wind speed sites in continental Europe become scarce, it is more economical to facilitate development in high wind areas on land, such as in the UK and Ireland, and offshore in the North Sea and Irish Sea. To facilitate this development, it would be necessary to plan for extensive new transmission in the North Sea, as well as between Ireland and the UK, in order to avail of this secure source of electricity.

A PROJECT FOR EUROPE: EUROPEAN-WIDE SUPERGRID

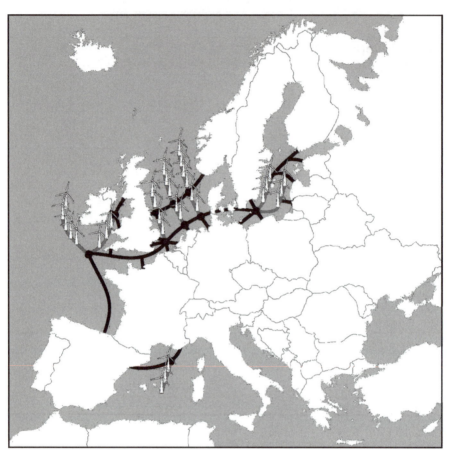

Source: Airtricity (2006)

Figure 10.8 *The Supergrid concept*

The Supergrid (see Figure 10.8) is a proposal for a high voltage sub-sea transmission network. It could, ultimately, cover the Baltic Sea, the North Sea, the Irish Sea, the English Channel, the Bay of Biscay and the Mediterranean. The Supergrid treats wind as a continental resource and would enable EU member states to share in this enormous energy source to their mutual advantage.

This could be achieved by the member states cooperating in the capture of their common wind resources and the conversion of this free energy into a reliable and predictable supply of electricity. Since power is always being generated on the Supergrid, it can be fed into the national grids to meet electricity demand.

The Supergrid also acts as an inter-connector between national markets and thereby helps to create a properly functioning internal market in electricity. This would bring additional benefits for European consumers in terms of greater competition, lower prices and increased security of supply.

The first steps to an EU SuperGrid: A 10GW North Sea project supplying the UK, The Netherlands and Germany

An initial analysis using a single year of historical wind data was undertaken in order to estimate the benefits to variability of such an offshore project (Hughes and Hurley, 2005b).

Generation of wind power production time series

Three wind power production centres surrounding the North Sea were chosen within the following countries (see Table 10.7):

1   Scotland, representing the majority of UK onshore capacity;
2   The Netherlands;
3   Northern Germany, representing the majority of Germany's onshore capacity.

Table 10.7 *Three wind power production centres surrounding the North Sea*

Locations	Latitude (°)	Longitude (°)	Capacity (MW)
Scotland (UK)	56.2	–2	1097
The Netherlands (NL)	52.2	5.7	1186
Northern Germany (DE)	52.2	9.5	17,500
North Sea (NS)	56.2	3.8	10,000

Source: Hughes and Hurley (2005b)

Production losses were not deducted since the stochastic events leading to most losses in a wind power time series are not readily modelled; furthermore, because this study aims to look at the frequency of changes in power production, the addition of losses adds an unnecessary layer of complexity to the results.

Allocation of North Sea power production

An initial goal was set to dispatch the power necessary to supply 100 per cent of load, when wind is available, directly to the UK and The Netherlands. The remaining production is absorbed by the large consumption centre of Germany. Table 10.8 outlines the resulting proportions allocated over the year to each country.

Table 10.8 *Proportions of annual North Sea power allocated to each country*

	Total share of North Sea power	Proportion (percentage)
Scotland (UK)	532,757	6.4
The Netherlands (NL)	874,976	10.6
Northern Germany (DE)	6,863,749	83
North Sea (NS)	8,271,482	100

Source: Hughes and Hurley (2005b)

In order to assess the variability reduction benefits of the addition of North Sea power, the resulting power series at each of the three locations (both with and without the North Sea power addition) were analysed in order to determine the frequency of changes in power production.

Figures 10.9 to 10.11 show power production changes between each six-hour interval.

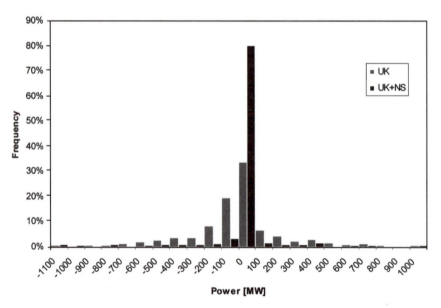

Source: Hughes and Hurley (2005b)

Figure 10.9 *The UK, with and without the North Sea*

It is clearly evident that the addition of the North Sea power to both the UK and The Netherlands could almost eliminate the variability entirely in two of the countries. Table 10.9 shows the proportion of the counts in the 0MW to 100MW 'bin' where there was no change in power over the six-hour interval.

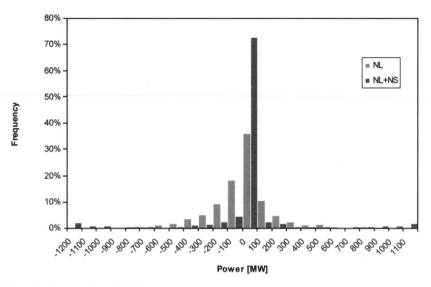

Source: Hughes and Hurley (2005b)

Figure 10.10 *The Netherlands, with and without the North Sea*

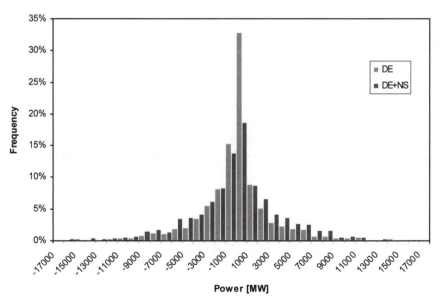

Source: Hughes and Hurley (2005b)

Figure 10.11 *Germany, with and without the North Sea*

Table 10.9 *Frequency of constant power in the UK and The Netherlands*

	Frequency of constant power (percentage)
UK	77
The Netherlands	69

Source: Hughes and Hurley (2005b)

This effectively firms up wind power to such an extent that it is much more reliable, especially if the variations in wind power can also be forecast.

This improvement is achieved with a much less pronounced effect on the German wind power series. The resulting variation is not markedly different from the scenario without the North Sea power. The noticeable changes are a reduction in small variations and a slight increase in medium-sized capacity changes.

The interconnected nature of the northern continental European grid may make the resulting German wind power variations more manageable than would be the case for an island system such as the UK, where the reduction in variability is a great benefit.

CONCLUSIONS

What emerges from the investigation is that when medium to high wind speed sites in continental Europe become scarce, it is more economical to facilitate development in high wind areas on land, such as in the UK and Ireland, and offshore in the North Sea and Irish Sea. To facilitate this development, it would be necessary to plan for extensive new transmission in the North Sea, as well as between Ireland and the UK, in order to avail of this secure source of electricity.

The Supergrid is a proposal for a high voltage sub-sea transmission network. It could, ultimately, cover the Baltic Sea, the North Sea, the Irish Sea, the English Channel, the Bay of Biscay and the Mediterranean. The Supergrid treats wind as a continental resource and would enable EU member states to share in this enormous energy source to their mutual advantage. This could be achieved by the member states cooperating in the capture of their common wind resources and the conversion of this free energy into a reliable and predictable supply of electricity. The Supergrid also acts as an inter-connector between national markets and thereby helps to create a properly functioning internal market in electricity. This would bring additional benefits for European consumers in terms of greater competition, lower prices and increased security of supply.

REFERENCES

Airtricity (2006) *European Offshore Supergrid Proposal: Vision and Executive Summary*, Airtricity, Dublin

Dowling, P., Hurley, B. and Giebel, G. (2004) *A Strategy for Locating the Least Cost Wind Energy Sites within an EU Electrical Load and Grid Infrastructure Perspective*, Proceedings of EWEA Conference, London

EWEA (European Wind Energy Association) (2004) *Wind Energy: The Facts – An Analysis of Wind Energy in the EU-25*, EWEA, Brussels, February

Garrad, A. D., Wastling, M. A., Quarton, D. C., Wei, J. (Garrad Hassan and Partners, Bristol), Matthies, H. G., Nath, C., Schellin, T. E. (Germanischer Lloyd, Hamburg), Scherweit, M. and Siebers, T. (Windtest KWK, Kaiser-Wilhelm-Koog) (1996) 'Study of offshore wind energy in the European Community', Garrad Hassan and Partners, Bristol

Giebel, G. (2001) *On the Benefits of Distributed Generation of Wind Energy in Europe*, PhD thesis, Carl von Ossietzky Universität Oldenburg, Fortschr.-Ber, VDI Reihe 6 no 444, Düsseldorf, Germany

Giebel, G., Kariniotakis, G. and Brownsword, R. (2003) 'The state of the art in short-term forecasting of wind power: A literature overview', Position paper for the Anemos project, http://anemos.cma.fr/download/ANEMOS_D1.1_StateOfTheArt_v1.1.pdf

Greenpeace (2004) *Sea Wind Europe*, Greenpeace, London, February

Hughes, P. and Hurley, B. (2005a) *Examination of the Benefits of Europe-wide Dispersion of Offshore Wind Power*, Internal Report, Airtricity, Dublin, December

Hughes, P. and Hurley, B. (2005b) *European Onshore Wind Variability Reduction through the Addition of 10GW of North Sea Production*, Internal Report, Airtricity, Dublin, December

Hurley, B. and Dodhia, K. (2006) *Review of Short Term Wind Power Forecasting Technology and Investigation of the Performance of Two Forecasting Models*, Internal report, Airtricity, Dublin, January

Joensen, A., Landberg, L. and Madsen, H. (1999) 'A new measure–correlate–predict approach for resource assessment', in *Proceedings of the European Wind Energy Conference*, 1–5 March, Nice, France, pp1157–1160

Kalney, E. et al (1996) 'The NCEP/NCAR 40-Year Reanalysis Project', *Bulletin of the American Meteorological Society*, vol 77, pp437–471, http://wesley.wwb.noaa.gov/reanalysis.html.

Landberg, L. (1997) 'The availability and variability of the European wind resource', *International Journal of Solar Energy*, vol 18, pp313–320

PB Power (2002) *ETSU Concept Study: Western Offshore Transmission Grid*, ETSU, London, February

11

Planning for Variability in the Longer Term: The Challenge of a Truly Sustainable Energy System

David Infield and Simon Watson

INTRODUCTION

Understandably, discussion of integrating wind energy and other variable renewables focuses on the electricity supply system. After all, wind and many of the renewables are, first and foremost, generators of electricity.

It seems clear that electricity systems can absorb low levels of renewable energy generation with little or no noticeable impact. Indeed, few would claim that the current levels of renewable electricity penetration of approximately 4 per cent in the UK (Janes, 2006) have any significant impact on the operation of the power system as a whole.[1] Even with 10 to 20 per cent penetration, integration costs are relatively minor and affordable, as is consistently stated in the preceding chapters of this volume.

Wind power, as the most developed of the new renewables, has been the main focus of discussion. The primary function of varying renewable sources such as wind is to displace generation from fossil-fuelled sources, thereby reducing carbon dioxide (CO_2) emissions; but all this book's authors agree that there will be a contribution to security of supply. At low penetrations, wind, for example, provides a capacity credit equal to its mean output (i.e. rated power multiplied by annual capacity factor). It is also agreed that as the penetration of wind (and this can be extended to other variable sources) increases, so the capacity contribution decreases, necessitating a greater plant margin to be held on the system in order to deliver the same degree of system security. This conclusion derives from the statistical characteristics of the resource availability and any correlation with loads. Whether the peak demands for electricity are more likely to occur on cold, windless anti-cyclonic days, or windy days with a strong wind chill factor, is an important question that requires further research. What is clear, though, is that introducing a wider diversity of renewable sources will ease the integration problem since the correlation between the different sources is generally weak or non-existent.

The purpose of this concluding chapter is to explore the longer-term issues of very high renewable energy penetration. Some of the earlier chapters, in particular Chapter 9 on a renewable electricity system for the UK, have prepared the way. It will be shown that an electricity supply system dominated by renewable energy cannot be viewed in isolation from wider energy supply and demand issues. Discussion here will focus on the UK; but the conclusions are likely to apply broadly to other national energy systems, particularly those with growing renewable energy penetration and liberalized electricity markets (e.g. within parts of Europe and across the US). If anything, the interconnected power systems found elsewhere in Europe make integration more straightforward.

RENEWABLE SOURCE-DOMINATED ENERGY SUPPLY SYSTEMS

It will be several decades at least before renewable energy becomes the dominant form of energy in the UK, and by that time a number of renewable energy technologies are expected to be established and competitive. These may include offshore wind, tidal stream energy, tidal barrages or lagoons, wave energy, photovoltaics (PV), and bio-generation. All of these except bio-generation are variable sources, although the nature of the variation and the extent to which they can be forecast differs significantly between the different sources.

Many global energy scenarios have suggested that renewable energy could become the dominant source of energy in the longer term, often viewed as beyond 2050. Shell strategists, in particular, have contemplated such a situation (Shell, 2001). They foresee a period in which renewable energy grows rapidly, and many would suggest that with wind seeing an annual global growth rate of over 25 per cent (Zervos, 2006) and PV growing at over 30 per cent per annum (admittedly from a low base), we have already entered this era. More importantly, the Shell strategists point to a possible slowdown in renewable energy uptake as integration issues start to cut in around 2030. It is certainly true that genuine integration issues exist. The need for increased margins is discussed in this book on several occasions, together with the importance of plant that can be controlled to provide system stability. Care must be taken, as highlighted by Chapter 6, to ensure that the drive to optimize the performance of individual plant items does not result in failure of the system as a whole. It is not apparent that the market mechanisms that currently determine the plant mix pay sufficient attention to such operational issues. Indeed, there is growing concern that the UK market does not provide sufficient incentives for the installation of new generation capacity despite the known projections for plant retirement.

Major changes have been seen over the last 20 years in the electricity market worldwide and in the UK in particular. Indeed, prior to the privatization of the UK electricity industry at the end of the 1980s, a true market did not exist. Instead, there was an electricity pricing system based on marginal costs with no real competition. The commercial development of wind energy since the 1980s and 1990s needs to be viewed against a backdrop of fundamental changes in the market system for pricing electricity. Some of these changes have been for the better as far as wind energy is concerned; the opening up of competition in the electricity supply market has given opportunities for niche players, such as green electricity companies, to enter the electricity retail arena. Other changes have not been so welcome. In particular, the UK's New Electricity Trading Arrangements (NETA), introduced in 2001, initially in England and Wales but now encompassing Scotland under the British Electricity Transmission and Trading Arrangements (BETTA), penalizes *individual* market players for their unforeseen electricity imbalances, whereas previously the cost of balancing supply and demand was spread out over all generators and suppliers. In the early days of NETA, as the new market was developing, this had a very negative impact on generators whose output could not be firmly predicted ahead of time – in particular, wind power and combined heat and power (CHP) (Bathurst and Strbac, 2001). Another of the great uncertainties with regard to UK wind farm development is the degree to which the transmission network operator, the National Grid, is willing or able to invest in new transmission line capacity. One of the keys to integrating wind power within a large-scale electricity grid is to ensure a geographically diverse generation base of wind farms (Halliday, 1988). Nevertheless, it makes sense to develop those areas with the best wind resource. These tend, however, not to be located in the most densely populated areas. In addition, much of the UK demand is concentrated in the south. To address this mismatch, new transmission capacity may be required if wind energy is to meet a substantial amount of the UK's electricity requirements, and this may necessitate a revision of the regulated market incentives for the transmission network operator.

Evolution of the electricity supply system has not been left entirely to the market. Regulation has an important role to play, and its importance may well increase. A good example of regulation in the area of power system operability is the National Grid's Grid Code,[2] which states that larger wind power installations, in common with other comparably sized generation plant, must be able to ride through faults and that, in the future, they must be required to contribute to frequency control and, thus, to overall system stability. In order to provide frequency control, wind turbines would need to be able to adjust their output either up or down in response to external signals, and this would require them to operate part loaded under normal operation. There is a cost to be paid for this, and whether it is the most cost-effective means of guaranteeing system stability is not clear. Nevertheless, similar requirements are likely to be introduced for other renewable energy sources, as and when their installed capacity reaches significant levels. Perhaps this could be seen as an opportunity rather than a threat to renewable energy generators. Participation in the balancing market in Great Britain can provide lucrative returns to flexible generators prepared to adjust their output at short notice. In the short term, while penetration levels for embedded intermittent generators remain low (meaning that they have little market leverage), this may not yet be practical, as discussed later in this chapter. Furthermore, effective operation of variable or intermittent renewable energy sources, such as wind farms, in such a market requires accurate forecasts of expected power output, although much work is being done in this area (see Chapter 5 on wind power forecasting). Indeed, studies indicate that forecasting could increase the value of wind power by up to 10 per cent (Giebel et al, 2003).

ENERGY STORAGE AND DEMAND-SIDE MANAGEMENT

A number of the chapter authors have referred to energy storage. Certainly, if cheap and effective energy storage were to become available, it would be widely used in electricity generation systems. This would be irrespective of renewable energy considerations, and would simply reduce the generation capacity required to meet time-varying demand with a given level of reliability. Indeed, if the storage could be small in scale and distributed, this would have the added benefit of reducing the capacity requirements of both the transmission and distribution systems.

Although energy storage technology has been widely discussed in the context of electricity supply systems (e.g. Black and Strbac, 2006; Barton and Infield, 2004), aside from large-scale pumped hydro systems, such as Dinorwig, that have the disadvantage of being geographically very specific, no cost-effective solutions currently exist. Target breakeven costs for energy storage for the UK electricity supply system have been estimated. Black and Strbac (2006) suggest a range of UK£252 per kilowatt (kW) to UK£970/kW based on operational savings and depending upon the plant mix; and this is significantly lower than even the optimistic projections for the latest technology of flow cells (Price et al, 1999). It is possible, however, that new and improved energy storage technology may achieve lower costs in the future. There is no shortage of advocates for different technologies, including compressed air, advanced batteries, super-conducting magnetic energy stores and flywheels. More recently, some have been proposing, and even demonstrating at a small scale, hydrogen-based storage systems, a subject to which we will return later. Overall, though, none of these currently provides a cost-effective solution for the electricity supply sector, which is why they are not in current commercial use.

Energy storage, however, is intrinsically associated with many end uses of electricity, as already mentioned in Chapter 9. Refrigerators and freezers, for example, have considerable thermal capacity and can maintain an acceptable temperature over many minutes without consuming electricity. Likewise, thermal mass in the fabric of buildings can allow heating and cooling plant to be rescheduled with few adverse impacts on comfort. Domestic hot water tanks can store 'electricity' effectively over hours with minimal loss. None of this is new, and, as already discussed, traditional off-peak electricity tariffs made use of this intrinsic storage to level out the daily electricity-demand profile. Recent technology developments (specifically, low-cost processing power and communication systems) have, however, opened up the possibility of further exploiting such opportunities, and currently there is considerable enthusiasm for new forms of demand-side management.

Traditionally, demand-side management has also embraced the use of commercial tariffs that allow disconnection of loads when generation capacity is limited. Although this can be useful to the system operator, there is a cost to pay. A company, for example, may incur production delays or other costs as a consequence of unplanned load shedding. In contrast, some of the ideas now being developed might be able to deliver changes to the consumption profile with no discernable negative consequences. 'Dynamic Demand' (see www.dynamicdemand.co.uk) is one such concept: in this approach, the compressor on/off temperature thresholds of fridges and freezers are adjusted to reflect the grid frequency. When the frequency is above the nominal UK network frequency of 50 hertz (Hz), indicating a transient surplus of national generation capacity over load, the switching set points are pushed up by an amount proportional to the instantaneous excess over 50Hz. This has the effect of making it more likely that the compressor is switched on and (on average, across many devices) ensures that the aggregate electrical load from such units increases. If the system frequency is lower than 50Hz, the reverse occurs, offloading the system. In this way, the automatic switching of the fridges and freezers contributes to frequency regulation, reducing the need for the system operator to hold plant on governor action for this purpose and thereby saving fossil fuel. Figure 11.1 shows how the compressor control depends upon system frequency with this approach.

Because the cost of mains frequency measurement is minimal and the dynamic demand control can be included in the more sophisticated controllers through simple adaptation of the control algorithms, these new appliances would be no more expensive to manufacture than the existing ones.

There is considerable commercial interest in the concept, and such appliances should benefit from the proposed Energy Efficiency Trading Scheme (EETS), which would allow some of the financial benefit to flow to the purchasers of these units. A preliminary study of the concept (Short et al, in press) has indicated that financial savings of up to UK£80 million per annum might be possible if this technology were widely taken up, most of this reflecting a reduction in conventional plant operating costs (largely fuel costs). There will be significant annual reduction in overall system CO_2 emissions as a result; but further research is required to quantify this. It is not known, however, whether there would be unwanted effects caused by this part of the national electricity load becoming more correlated, thus reducing demand diversity. Further research is required to examine this and to more accurately quantify the costs and benefits of the approach.

Although the Dynamic Demand concept was developed specifically for fridges and freezers, it could be applied to any other applications with intrinsic thermal storage and temperature control.

A larger renewable energy penetration will require a more flexible system. Adjustable loads could be just as important as flexible plant, and potentially much more cost-effective and environmentally sound. In the initial study of the Dynamic Demand concept (see www.dynamicdemand.co.uk), its potential to ease the integration of wind power was provisionally assessed. Figure 11.2 is taken from this study and shows that up to 13.8 gigawatts (GW) of wind

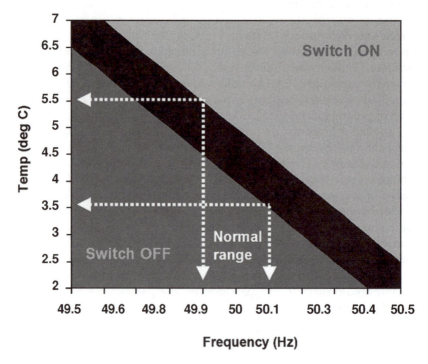

Figure 11.1 *Operation of the Dynamic Demand concept*

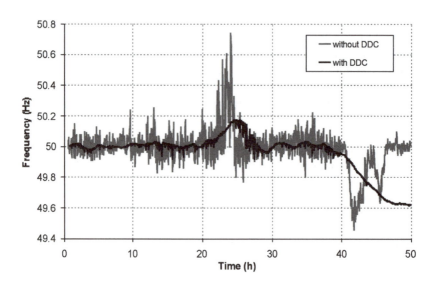

Figure 11.2 *Impact of Dynamic Demand control (DDC) on grid frequency with 13.8GW of wind capacity*

energy capacity can be absorbed without the need for additional frequency regulation.

Barrett's work in progress, reported in Chapter 9 of this volume, suggests that up to 95 per cent of the UK's electricity supply could be met from renewable sources. Similar conclusions were derived in research completed at CREST by Streater (2002), which also used hour-by-hour modelling of the aggregated UK electricity load and renewable supplies. Both studies ignored the detail of geographical distribution and related connection and transmission issues, although the positive contribution from the geographical diversity of the renewable energy supplies has been incorporated, to an extent, within Streater's (2002) study. There are some significant differences in the assumptions, most notably the inclusion of CHP in Barrett's study.

Figure 11.3 is reproduced from Streater (2002) and shows the hour-by-hour contributions from the different renewable energy sources and their aggregate compared with both the electricity (power) load and the aggregate heat and electricity load. The term 'source-load' is the surplus of supply over aggregate load and shows that for only about 12 weeks in the year is there an energy shortage, and also that for large parts of the year there is a significant energy surplus. In Barrett's model, this surplus is stored for future use. Whether this is attractive will depend upon the cost of energy storage; it may well be cheaper and simpler, overall, to curtail the output from the renewable devices. Note that in Streater's (2002) work, the combined renewable output far exceeds the electrical power demand throughout the year.

Of course, it may well be that energy storage appears in the system for other reasons. As discussed by Barrett (see Chapter 9), it may well be that a significant proportion of transport needs will be provided by electric vehicles. Whether these are powered by batteries or fuel cells is immaterial if the energy is supplied by the electricity system. In both cases, substantial energy storage is involved and this can be used to improve the utilization of the renewable resources. In fact, this storage may be essential for stable dynamic operation of the electricity supply system, and this critical issue is not addressed in either Barrett's (see Chapter 9) or Streater's (2002) studies.

Micro-generation (i.e. very small, ideally renewable, source-driven generators located on consumers' premises) is of increasing interest. How this develops in practice is currently unclear and will depend upon the economics relative to larger renewable energy installations, such as wind farms. This development may well ease the technical electrical integration issues since the generation will tend to be nearer to the loads; but it should not substantially affect the aggregated results presented here.

One of the challenges in delivering the developments discussed – in particular, extensive demand-side management and high levels of distributed generation – is to reward the participants. This may require adaptation of the existing market system. The present arrangements under BETTA allow for a balancing market where flexible generators and suppliers can bid/offer to change their positions at short notice. As mentioned above, this can provide significant financial returns to those players in the market who have sufficient flexibility. However, playing such a market needs investment in the appropriate

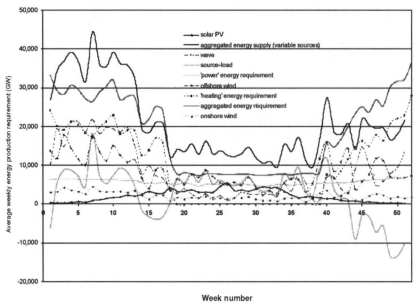

Source: adapted from Streater (2002)

Figure 11.3 *Hour-by-hour variation in renewable energy generation over one year, compared with variations in energy requirements*

information technology (IT) systems and dedicated trading teams. The present balancing market in the UK was designed for the major market players who can afford such investment. Smaller players, such as householders, small commercial companies and small embedded generators are not able to participate in such a market. First, they are at a level where their consumption is not metered half hourly; and, second, the value of adjusting their output individually would be less than the cost of the metering and associated IT systems that would be required for active participation. Besides, few small consumers or generators would have the time or resources to play the balancing market. The system would need to be changed substantially to allow household-level demand-side management and embedded generation to benefit from such a market. This would require low-cost smart metering and automatic systems driven by market price signals, which would necessitate little or no user intervention, possibly coordinated by the contracted electricity supply company. The supply company could then aggregate the benefits for a number of small consumers or generators and pass on the associated benefits. Such changes could open up a significant market for small-scale flexible demand and generation, and provide further flexibility to the system operator, which could aid the process of integrating variable renewable energy sources within the grid.

CONCLUSIONS AND FURTHER RESEARCH

To understand the impact of variable renewable energy generation on a large scale, a electricity network such as that within the UK requires detailed numerical modelling. A substantial amount of work has already been done in this area. However, much of this has been based on older market systems at a time when there was no significant wind power on the network. Further work is required to quantify future load margin requirements, plant cycling, and so on, particularly in light of possible changes in demand patterns driven by climate change. Many of the earlier analyses of network impact were based on hypothetical wind power generation; now that data exists for actual wind power generation, it should be possible to refine and expand existing network models. Further work is also required to model the effect of demand-side management, including smart metering, smart appliances and small-scale embedded generation. Equally importantly, expected changes in future consumer behaviour, and what impact such changes are likely to have on future demand profiles and the uptake of demand-side management, must also be quantified. These different aspects require not only technical studies of how networks are able to accommodate intermittent generation, but should also consider socio-economic factors, particularly with regard to consumer behaviour.

The challenge of integrating intermittent renewable energy generation should not only rely on technical fixes, but will also require a significant change in public behaviour. There is a need to move away from the traditional view of an electricity industry slavishly providing energy to passive consumers. Today's consumers are becoming, and will need increasingly to become, active participants in a modern sophisticated electricity network that needs to meet the twin challenges of providing electricity at an economic cost, but in a way that is sustainable. Technology can facilitate this process; but it is we, as consumers, who must embrace this new age and take responsibility for the energy that we use. In an age where human activity is now seen to be driving climate change, the incentive to move to a sustainable electricity industry with increasing reliance on renewable energy generation becomes ever stronger.

NOTES

1   It is important to distinguish between this and grid connection issues that may well require local reinforcement of the electricity distribution system.
2   This code is designed to permit the development, maintenance and operation of an efficient, coordinated and economical system for the transmission of electricity; to facilitate competition in the generation and supply of electricity; and to promote the security and efficiency of the power system, as a whole. The National Grid and users of the transmission system are required by law to comply with this. The Grid Code was last updated on 1 January 2007, and the current version is available at www.nationalgrid.com/uk/Electricity/Codes/gridcode/.

REFERENCES

Barton, J. P. and Infield, D. G. (2004) 'Energy storage and its use with intermittent renewable energy', *IEEE Transactions on Energy Conversion*, vol 19, no 2, June, pp441–448

Bathurst, G. and Strbac, G. (2001) 'The value of intermittent renewable sources in the first week of NETA', *Tyndall Briefing Note No 2*, Tyndall Centre for Climate Change, Norwich

Black, M. and Strbac, G. (2006) 'Value of storage in providing balancing services for electricity generation systems with high wind penetration,' *Journal of Power Sources*, vol 162, no 2, pp949–953

Giebel, G., Brownsword, R. and Kariniotakis, G. (2003) *The State of the Art in Short-Term Prediction of Wind Power: A Literature Overview*, Deliverable Report D1.1 from EU Project ANEMOS, Contract no ENK5-CT-2002-00665

Halliday, J. A. (1988) *Wind Meteorology and the Integration of Electricity Generated by Wind Turbines*, PhD thesis, University of Strathclyde, Rutherford Appleton Laboratory, UK

Janes, M. (ed) (2006) *Digest of United Kingdom Energy Statistics*, Stationery Office, UK

Price, A., Bartley, S., Male, S. and Cooley, G. (1999) 'A novel approach to utility scale energy storage (regenerative fuel cells)', *Power Engineering Journal*, vol 13, no 3, pp122–129

Shell (2001) *Energy Needs, Choices and Possibilities: Scenarios to 2050*, Shell International report, The Hague

Short, J., Infield, D. G. and Freris, L. L. (in press) 'Stabilization of grid frequency through dynamic demand control', *IEEE Transactions on Power Systems*

Streater, C. J. M. (2002) *Scenarios for Supply of 100% of UK Energy Requirements from Renewable Sources*, REST MSc thesis, Department of Electronic and Electrical Engineering, Loughborough University, UK

Zervos, A. (2006) 'Wind power development in Europe: Achievements and prospects', in *Proceedings of the European Wind Energy Conference*, 28 February–2 March, Athens, Greece

Index